Kaizen Kanban

Also available from ASQ Quality Press:

Making Change in Complex Organizations
George K. Strodtbeck III

The Strategic Knowledge Management Handbook: Driving Business Results by Making Tacit Knowledge Explicit
Arun Hariharan

Failure Mode and Effects Analysis (FMEA) for Small Business Owners and Non-Engineers: Determining and Preventing What Can Go Wrong
Marcia M. Weeden

The Quality Toolbox, Second Edition
Nancy R. Tague

Root Cause Analysis: Simplified Tools and Techniques, Second Edition
Bjørn Andersen and Tom Fagerhaug

The Certified Six Sigma Green Belt Handbook, Second Edition
Roderick A. Munro, Govindarajan Ramu, and Daniel J. Zrymiak

The Certified Manager of Quality/Organizational Excellence Handbook, Fourth Edition
Russell T. Westcott, editor

The Certified Six Sigma Black Belt Handbook, Second Edition
T.M. Kubiak and Donald W. Benbow

The ASQ Auditing Handbook, Fourth Edition
J.P. Russell, editor

The ASQ Quality Improvement Pocket Guide: Basic History, Concepts, Tools, and Relationships
Grace L. Duffy, editor

Office Kaizen: Transforming Office Operations into a Strategic Competitive Advantage
William Lareau

Modular Kaizen: Continuous and Breakthrough Improvement
Grace L. Duffy

To request a complimentary catalog of ASQ Quality Press publications, call 800-248-1946, or visit our website at http://www.asq.org/quality-press.

Kaizen Kanban

A Visual Facilitation Approach to Create Prioritized Project Pipelines

Fabrice Bouchereau

ASQ Quality Press
Milwaukee, Wisconsin

American Society for Quality, Quality Press, Milwaukee 53203
© 2017 by ASQ. Printed in 2016
All rights reserved.
Printed in the United States of America
21 20 19 18 17 16   LS   5 4 3 2 1

Library of Congress Cataloging-in-Publication Data
Names: Bouchereau, Fabrice, 1974– author.
Title: Kaizen kanban : a visual facilitation approach to create prioritized
    project pipelines / Fabrice Bouchereau.
Description: Milwaukee, Wisconsin : ASQ Quality Press, [2016] | Includes
    bibliographical references and index.
Identifiers: LCCN 2016021635 | ISBN 9780873899376 (hardcover : alk. paper)
    ISBN 9781636941455 (paperback)
Subjects: LCSH: Continuous improvement process. | Project management—Quality
    control.
Classification: LCC TS183 .B68 2016 | DDC 658.5/33—dc23
LC record available at https://lccn.loc.gov/2016021635

ASQ Mission: The American Society for Quality advances individual, organizational, and community excellence worldwide through learning, quality improvement, and knowledge exchange.

Attention Bookstores, Wholesalers, Schools, and Corporations: ASQ Quality Press books, video, audio, and software are available at quantity discounts with bulk purchases for business, educational, or instructional use. For information, please contact ASQ Quality Press at 800-248-1946, or write to ASQ Quality Press, P.O. Box 3005, Milwaukee, WI 53201–3005.

To place orders or to request a free copy of the ASQ Quality Press Publications Catalog, visit our website at http://www.asq.org/quality-press.

Quality Press
600 N. Plankinton Ave.
Milwaukee, WI 53203-2914
Email: books@asq.org

ASQ   Excellence Through Quality™

For Julia, my wife

Table of Contents

List of Figures and Tables

Part II

Foreword

It was Joseph Juran who said that "all improvement happens project by project and in no other way" (1989). This concept is particularly relevant to the methods discussed in *Kaizen Kanban*, in which Fabrice Bouchereau intricately weaves together traditional quality tools in an innovative way. This approach not only supports process improvement but also identifies problem areas that can be addressed through improvement projects.

My experience over nearly 20 years in the quality field is that many organizations struggle with identifying issues to address through improvement projects. They often want to improve their operations, but they don't know where to start. This is somewhat surprising given the plethora of problems that many organizations face; yet, it is likely that the sheer number of problems is what makes it hard for organizations to prioritize and select the ones most in need of attention.

The approaches described in *Kaizen Kanban* provide unique solutions to this problem. It focuses on helping organizations identify opportunities for improvement in an efficient and value-added way. The result is a prioritized project pipeline that was identified by leveraging the expertise of organization members using approaches built on well-known quality tools and methods, such as process mapping and failure modes and effects analysis (FMEA).

The power behind the kaizen kanban approach emanates from Bouchereau's innovative adaptations of traditional quality tools and methods. For example, concepts derived from SIPOC (suppliers, inputs, process, outputs, customers) diagrams and FMEA are integrated within process mapping activities, which significantly increases the depth of analysis within the process being examined. Then, logical and well-known approaches are employed to (1) prioritize the opportunities for improvement identified and (2) categorize them within the appropriate type of kaizen event. In the end, the kaizen kanban approach provides an effective method for coordinating project selection that efficiently feeds the deployment of improvement projects across an organization using a visual communication approach.

Those within various levels of an organization, from supervisors and managers through executives, in a wide range of industries will find the concepts and approaches discussed in *Kaizen Kanban* useful to drive performance improvement. The practical guidance provided will help strengthen the facilitation and problem-solving skills of organizational members. In so doing, organizations can transform themselves from firefighters and reactionaries to proactive identifiers of opportunities for improvement and effective problem solvers.

Jamison V. Kovach, PhD
Associate Professor and Director, Lean Six Sigma Professional Training Program
University of Houston

Preface

Welcome to *Kaizen Kanban*, a guide to creating prioritized project pipelines and setting up improvement boards to maximize business success through the execution of continuous improvement projects.

In this book you will be introduced to the "Faster and Better" visual facilitation approach that enables you to seamlessly leverage and combine fundamental tools in order to identify improvement opportunities for entire value streams, compile them in a prioritized project pipeline, and set up improvement display boards, or "kaizen kanbans," linked to key business objectives.

This approach is designed to complement and enhance the effectiveness of quality, lean, continuous improvement, and project management initiatives that may already be in place in an organization.

Acknowledgments

I would especially like to thank:

- My parents, Jean-Marie and Magaly Bouchereau, who taught me to always strive for faster and better successes

- My brother Raphael and my sister Valerie for always being there for me

- Cammy and Harris Tran for showing me how much you believed I could finish this project

- Ade Garcia and Bernard Vivy for listening to my concepts and providing priceless feedback

- Derek Smith for trying out my ideas during his facilitation sessions

- Bill Weller for his constructive criticism and his support during the writing process

- Eddie Merla for encouraging me to write this book and setting me up for success

- Keiko and Miso Tran-Bouchereau, who were literally at my side as I typed each word of this book

Special thanks to Matt Meinholz and Paul O'Mara at ASQ Quality Press for their support throughout the preparation and publication of this book.

Introduction

When I first started my career as an industrial and systems engineer, I dreamed of making an impact, of working on major projects that would change the course of companies and elevate organizations to the next levels of safety, quality, and profitability.

But things did not go exactly as planned. Like many of my peers involved in change management, I became a firefighter. I developed skills that allowed me to quickly react and address the burning issues of the day. For years I enjoyed being a superhero and saving products, processes, and people. Being a firefighter was fun; it brought instant gratification and gave me incredible stories I used to impress my family, coworkers, and potential next bosses.

I have now been involved with change for close to 20 years, and over time my responsibilities fortunately have become less about reacting to crisis and more about being proactive and strategic in nature. During this time I have observed numerous well-branded organizations invest large amounts of time and money in training initiatives to equip employees with the tools necessary to become effective change agents.

Most of these organizations tended to favor mainstream methodologies such as lean, Six Sigma, and project management, which complement and reinforce classroom learning by requiring all participants to apply their newly acquired skills to actual projects.

I strongly believe completing projects is fundamental in developing a student's ability to understand and apply lean, Six Sigma, and continuous improvement methodologies. However, I have frequently noticed that participants in these programs struggle to complete their first project just as I did and often for the same reasons.

I was first exposed to Six Sigma in 2001 in a mandatory introductory course offered by my employer at the time. Three days before the start of class, I was asked to identify and charter two projects as part of my Green Belt certification prework. I had no idea what a project charter was, and I arrived at the class empty-handed, fearing I was in trouble and feeling very inadequate.

A few years later I accepted a position at another company and found myself in a similar situation. Two weeks before my start date, while on the road moving cross country, I received an e-mail informing me that I had been registered for a Black Belt and that I was expected to have two Black Belt projects chartered within a week. How could the company expect me to identify opportunities for improvement if I had never set foot in the plant?

Since then I have witnessed countless organizations that had the same expectations of employees about to start their training. Invariably these candidates felt set up for failure and often arrived to class with a high level of anxiety and even resentment.

Countless hours observing smart and motivated trainees stumble as they kicked off their first project revealed two main contributors to this situation: poor project selection and a lack of facilitation skills.

This realization led me to develop the "Faster and Better way" (FAB~DO)[1] to address these gaps by complementing traditional problem-solving methodologies with a step-by-step facilitation approach to identify improvement opportunities for entire value streams, compile them in a prioritized project pipeline, and set up improvement display boards, or "kaizen kanbans," that are linked to key business objectives and visible to everyone.

NOTE

1. I substituted "DO" for "way" in reverence to my Japanese senseis who started me on the way.

Part I
Prioritized Project Pipeline

The range of what we think and do is limited by what we fail to notice. And because we fail to notice that we fail to notice there is little we can do to change until we notice how failing to notice shapes our thoughts and deeds.

—R. D. Laing

1

Why Kaizen Kanban?

The Japanese word *kaizen* is commonly used to describe a team approach to quickly break apart a process and rebuild it in order to function better. It is a philosophy that advocates continual process improvement. In this book, we will discuss the five types of kaizen that can be used at different levels of the organization by practitioners as they achieve new skill levels.

Kanban is the Japanese word for "signboard" or "billboard." In the traditional approach, kanban cards are used to signal to workers what to build next or what parts to retrieve.

Kaizen kanbans, or improvement display boards, follow the same principles used with traditional kanbans. They are visual communication tools and are visible to all levels of employees within the organization.

The difference is that instead of telling operators what to build next or what parts to retrieve, the cards tell improvement teams what pre-approved projects are most relevant to current business needs and are next in line for implementation.

2
Context

In most companies, the day-to-day priorities of keeping the production line running, meeting shipment commitments, and interfacing with suppliers and customers are urgent matters that need to be dealt with immediately.

A change agent is like a catalyst. A change agent facilitates the collaboration between two or more individuals or groups by creating conditions that empower them.

Quality improvement professionals and project managers have a wealth of knowledge and a plethora of tools that can be used to streamline and improve processes. They should be change agents, but unfortunately most of their time is spent doing routine work and "fighting fires."

In this book, we will use a case study to retrace the steps of a change agent named Bill D. Smith and discover how he combined fundamental tools and techniques from various toolboxes into a coherent, dynamic, and customizable improvement strategy to create a kaizen kanban for an organization called SportsMax.

3

SportsMax

SportsMax is a large Texas corporation that manufactures and sells recreational and sporting equipment. SportsMax's growth over the past decade has come mostly through acquisitions.

SportsMax recently decided to expand its product line and purchased Velocycle, a small manufacturer and distributor of bicycles located in Miami, Florida. Velocycle has been in business and profitable for more than 20 years. Despite enjoying a great reputation and having a loyal customer base, Velocycle always seemed to lack the resources to grow and expand.

SportsMax CEO Maggie Jean believes the new acquisition has the potential to become a very profitable product line and has decided to send a team of six highly performing SportsMax employees to evaluate the current business practices at Velocycle:

- Garcia, health and safety manager

- Jason, manufacturing engineering manager

- Ben, supplier development engineer

- Camille, quality and documentation expert

- Rafi, sales manager

- Valerie Lee, financial analyst

4

Findings and Report

After a week in Florida, the team members returned to Texas and met with the CEO to discuss their observations. The list was extensive and has been summarized here:

- No health and safety manager on site
- No evidence of a safety plan
- Poor ergonomic practices
- Nonconforming material mixed with good parts
- Work-in-process inventory accumulated at work stations
- Out-of-calibration measurement instruments
- Poorly documented material specifications that frequently resulted in incorrect materials being purchased
- Finished goods that could not be shipped because the right assortment of sizes and colors was not available
- Large stacks of unpaid bills and overdue invoices

These bad management practices, along with a lack of familiarity with Velocycle processes, led the SportsMax management team to hire Bill Smith. He was tasked with assessing what it would take to turn the company around in three months or less and creating an action item list.

Bill Smith is an industrial engineer who specializes in company-wide lean and quality transformations. His success is often attributed to his team management and facilitation style.

His unique approach to facilitation leverages best practices and minimizes the unnecessary customization of well-known tools. This enables him to combine various toolboxes and the experience of participants to quickly identify improvement opportunities for entire value streams.

5

Prework

Bill flew to Miami, took a quick tour of the plant, and met with key stakeholders to better understand their expectations.

> **Stakeholders:** People affected by the outcome of the project; they may influence results but are not necessarily directly involved with project work.

Nayla Conrads, general manager of the plant, and the executive staff of Velocycle were identified as key stakeholders. They prepared the initial draft of the project charter. This formal document set the objectives to achieve and gave Bill the authority he needed to obtain resources and assemble a team.

> **Project charter:** A tool used to concisely describe the work of the team and communicate the purpose of the project to all attendees.

Bill favors the simplicity of one-pagers that incorporate the flexibility to tailor the charter to the needs of the project. Knowing that templates are widely available and come in a vast array of formats and complexity, Bill made sure the Velocycle management team included the following information (Figure 5.1):

- Project name
- Schedule
- Champion's name
- Facilitator's name
- Problem statement
- Business case
- Project description

Project name:	Velocycle Modernization and Integration
Schedule:	July 1–5
Champion:	Nayla Conrads, general manager at Velocycle
Facilitator:	Bill D. Smith
Problem statement:	The Velocycle production process consists of steps executed in a series or in parallel by team members working for different departments. Currently there is not a clear understanding or agreement of what is expected of those who work in the various functions, nor is there agreement on the tools they use.
Business case:	The Velocycle team struggles to keep track of inventory, has a large number of reworks, and frequently fails to deliver on time. The new managment team anticipates a surge in demand as the economy recovers from the downturn. With competitors continually striving to gain market position, Velocycle must continue to reinvent itself, meet customers' expectations, and increase profitability.
Project description:	Review current practices and systems used at Velocycle and identify opportunities for improvement.

Figure 5.1   Velocycle modernization project charter.

WALKING THE PROCESS

Once the charter was finished, Bill spent the next few days on the production floor. He observed the process and spoke with workers from the various departments.

He learned that both new and longtime operators feared losing their jobs as a consequence of the acquisition by SportsMax. They had heard and believed a rumor that SportsMax owned other bicycle manufacturing plants and would close the Miami location to eliminate duplicate staff.

He immediately addressed their concerns and alleviated their fears by explaining that there were no other plants. He informed them of the plan to take Velocycle to the next level. After hearing this, the newer operators immediately wanted to make improvement suggestions. More experienced operators, however, were noticeably less eager.

A particularly important factor in change management success is to gain the buy-in of employees who are skeptical because of previous initiatives that yielded few results or because the company failed to implement or sustain their recommendations.

Bill conducted interviews and administered surveys to identify the tools used by the workers at the various operations, the challenges they faced, and the opportunities they would like the team to look into.

He obtained copies of local work instructions, standard operating procedures (SOPs), and existing process maps from the operators, who also volunteered a large amount of tribal knowledge.

Standard operating procedure (SOP): A policy and procedure document that describes the regular recurring activities appropriate to quality operations.

> **Tribal knowledge:** Undocumented information that is assumed to be factual, but without proof, and is handed down from one generation to the next within a group.

HIGH-LEVEL PROCESS BREAKDOWN

Bill used his newly acquired insight and process knowledge to create a high-level process breakdown as a starting point for scoping at the kickoff meeting.

He captured the process from receiving to shipping in six steps:

- Document process
- Make frame
- Paint frame
- Install components
- Install wheels
- Pack bicycle

> When possible, the facilitator should attempt to break down the process into no more than six high-level steps, written in the verb-noun format, to be used as a starting point for confirming the scope of the initiative at the kickoff meeting with the team.

Bill leveraged this high-level process breakdown to identify, compile, and organize all the related documentation his team would need to analyze the process.

AGENDA

Bill decided to schedule a multiday work session to further identify and prioritize the challenges faced by each department at Velocycle. He invited a cross-functional team and requested that several subject matter experts (SMEs) who could not be present for the entire meeting remain on call and available to participate.

> **Cross-functional team:** A group made up of people from different departments or who perform different functions within a company. Such teams are particularly useful when trying to solve problems or exploring potential solutions.

> **Subject matter expert:** A person who knows exactly what it takes to do a particular job and is considered to be an authority on a specific subject or area.

For a multiday session, having an agenda for each day is key to staying on track and meeting all the objectives.

> **Agenda:** A list of topics or points to discuss during a meeting, workshop, or work session.

Bill used the agenda to set objectives for each day and communicate them to the team in a clear and concise manner to maintain a sense of urgency and enable the timer to keep track of progress.

TEAM ACTIVITY PROGRESS TRACKER

Bill specified the tangible deliverables for each item on the agenda and incorporated them into an easy-to-follow rolling action item list (RAIL) so he could monitor progress and maintain a sense of urgency for all team members (Figure 5.2).

#	Category	Action	Owner	Start date	Target date	Status	Priority

Figure 5.2   Rolling action item list.

The easiest way to set up a RAIL is to draw it on a flip chart, place it on a wall, and use it like a to-do list, where the status of each item (on time, late, or complete) can easily be updated.

Rolling action item list (RAIL): A tool used to keep track of the actions the team is set to accomplish and their status.

By taking the time to translate the team deliverables into an easy-to-follow RAIL, the team can constantly measure success, maintain a sense of urgency, and adjust the level of intensity of efforts as needed.

TEAM SELECTION

Bill knew that having the right team members would be a significant factor for success, so he carefully chose them based on their process knowledge, their positive attitude, their willingness to share and to learn, and their eagerness to try new things.

The optimal number of direct team members varies depending on the scope of work. The facilitator has to balance between choosing too large a number, which could result in chaos, and choosing so few participants that they cannot generate enough ideas.

Bill is a firm believer that process improvement tools are most effective when used in a team setting. He shared with the team that, as a facilitator, he always made sure to have a cross-functional representation of SMEs on hand if possible, because cross-functional teams tend to accomplish more in a short period than regular teams.

Cross-functional teams are made up of employees at about the same hierarchical level but from different work areas who come together to accomplish a task. Cross-functional teams are an effective means of allowing people from diverse areas within an organization to exchange information, develop ideas, solve problems, and coordinate complex projects. (Manos and Vincent 2012)

When selecting team members, it is a good practice to seek personnel familiar with the area of the organization where the problem was observed, as well as where the problem may have originated. SMEs may also contribute valuable insight.

It is equally important to include individuals with limited exposure to the processes to be investigated, as they may question "the obvious," to which the other participants who see it every day have become blind.

TEAM ROSTER

Over the years, Bill adopted the best practice of specifying the role of each person on the team, as it makes it easier for all team members to know who to contact if they need more information (Figure 5.3). For example, somebody may currently hold the title of receiving dock manager but be on the team as the documentation expert because they were in charge of compliance in their last position.

First name	Last name	Department	Manager	Role on this team

Figure 5.3 Roster template.

6
Kickoff Meeting

With all the prework completed and the team in place, it was finally time for the kickoff meeting.

Kickoff meeting: The first meeting with the project team and the client of the project.

It is the ideal opportunity to emphasize the importance of:

- Confidentiality: "Our discussions remain in this room"
- Focusing on the task at hand: "If you had something more important to do, you would not be here right now"
- Getting a good return investment for our time and effort

The kickoff meeting is usually the first time the members of the team officially meet to work on a particular project. They may or may not have collaborated in the past; however, this is the first time they will be collaborating on this task and should take the time to learn why each person was selected.

The first contact with the team was the ideal opportunity for Bill to set the tone for all future team interactions. He was fully aware that the team would go through the team life cycle of forming, storming, norming, performing, and disbanding. But he hoped that the tools and approaches he would be leveraging would be effective at moving the team from the forming to the performing stage as fast as possible.

During the kickoff meeting, Bill distributed the prework information package he had compiled for the team, reviewed the agenda, and set the ground rules.

GROUND RULES

Bill set ground rules with the team early in the process to set the tone and secure the buy-in of all participants. This provided him and the team members a road map for handling crises and reducing the likelihood of conflict.

Ground rules: Rules that are generic in nature and promote the creation of a serene and peaceful atmosphere, which encourages creativity.

Good ground rules take into account that people have responsibilities and needs beyond the scope of their involvement with a particular initiative.

17

For short sessions, the facilitator can follow a standard list of generally accepted meeting ground rules to set the tone and get the meeting moving. For longer sessions or workshops, the facilitator may encourage the team to create its own meeting ground rules and capture them on a flip chart so that they can be on display during the entire workshop.

Examples of session ground rules include the following:

- Be kind and considerate

- Turn off computers and silence cell phones

- Excuse yourself and take urgent calls outside the room

- Enter into discussions enthusiastically

- Give freely of your experience

- Allow and encourage others to contribute

- Ask questions when you do not understand

- Appreciate the other person's point of view

- Provide constructive feedback—and receive it willingly

- Confine your discussions to the topic

- Keep things straightforward and simple

- Be innovative—encourage creative solutions

- Most of all, have fun!

PARTICIPANT INTRODUCTION

Bill started the meeting by requesting that the participants introduce themselves. He displayed the template shown in Figure 6.1, prompting the participants to share the following information with the team:

- *Name:* People who are sensitive about the pronunciation of their name can use this opportunity to let the team know how to address them.

- *Job title:* This may help the team understand the perspective from which the individual views the world. A safety manager will look at the world from a different set of lenses than that of a quality manager.

- *How long you have been with the company:* A longtime employee may bring historical information that could help the team avoid repeating mistakes of the past, while a brand-new employee may bring experience acquired at other workplaces that the team can use for benchmarking. Participants may also bring preconceptions that will have to be addressed at the appropriate time.

- *Your favorite inventor or invention (or favorite actor/sports team/movie):* This question is used as an icebreaker.

- *Why you are here:* This forces the person to share with the team their reason for being on the team, their "what's in it for me" (WIFM), and why they care about this project.

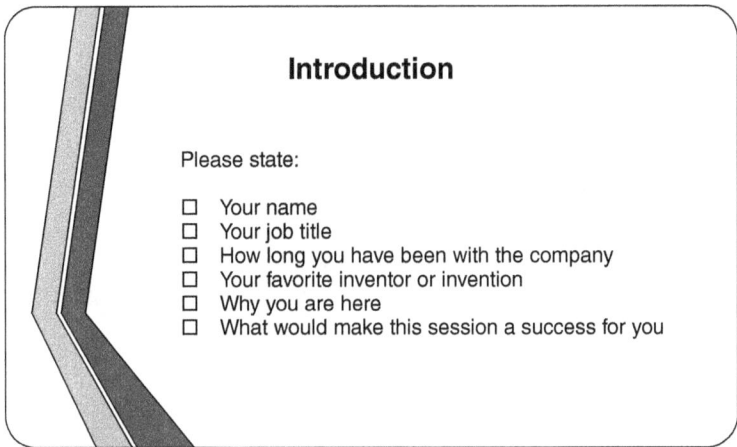

Figure 6.1 Introduction slide.

- *What would make this session a success for you:* This may be one of the most important bits of information obtained from this exercise, as it lets the other team members know what the individual is working toward. Record these statements of success on a flip chart and use them to align and get buy-in from the team members.

Getting to know the participants is very important. Understanding why they are here, what drives them, their preferred communication style, the role they play on the team, what they bring to the table, and how they define success for this task provides the facilitator with the insight needed to best align interests and avoid conflict.

REVIEW THE HIGH-LEVEL PROCESS MAP

With limited resources and an ambitious goal, Bill had to make sure that all the participants, many of whom came from different departments and had different backgrounds, were looking at the process. Together they reviewed the high-level process breakdown prepared by the champion during the prework and reformatted it to make the sequence and flow more obvious (Figure 6.2).

Champion: An executive-level manager who is responsible for managing and guiding the team for a particular initiative.

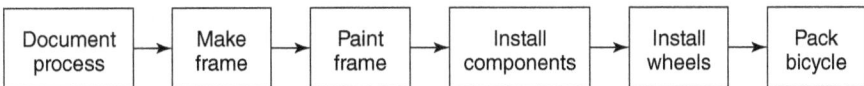

Figure 6.2 High-level process map.

IN SCOPE/OUT OF SCOPE

The team assembled to take Velocycle to the next level had to first focus all its efforts on identifying opportunities for improvement. Bill knew his team could not afford to be sidetracked, and the only way to prevent this was to have the team specify what aspects within the defined process would be included in the analysis. The team proceeded to scope the project.

They defined the start and end points of the high-level process map (think of it as identifying the boundaries between which the project will be contained). In Figure 6.3, everything green is in scope and can be considered; everything red is out of scope and will not be considered.

Figure 6.3   In-scope vs. out-of-scope activities.

> The start and end points should be tangible events or actions that make obvious the first and last steps of the process that the team will be analyzing.

They specified what aspects within the defined process (in green) would be included and what aspects should be excluded from the analysis.

Bill and his team decided to limit their analysis to the manufacturing process at Velocycle. They defined the arrival of raw material at the receiving dock as the starting point and the placement of finished material on a truck as the ending point for the manufacturing process (Figure 6.4).

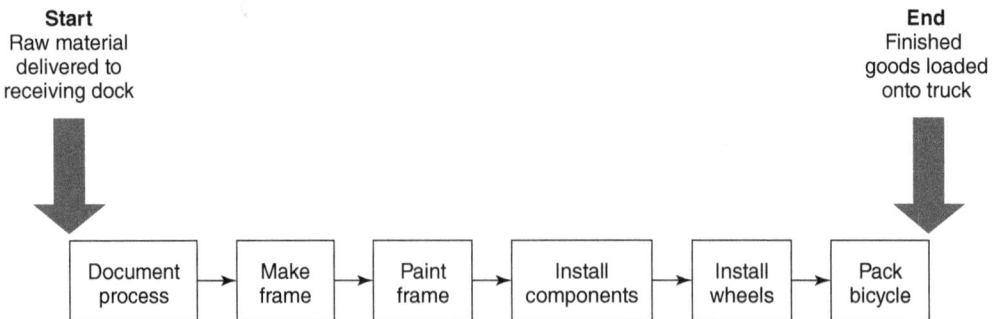

Figure 6.4   In/out scope for Velocycle manufacturing process.

7

Process Map

Once the scope of the project was agreed on, Bill had the team start creating a process map. He wanted a shared visualization of the process steps for his team that would help them gain a better understanding of how the system actually worked.

He presented process mapping as the initial step toward improvement because it gathers insight from those who work with the process. He insisted that the activity would lead the team to:

- Review the steps of the process and their sequence

- See non-value-creating steps or duplicated efforts

- Clarify relationships between people and the organization

- Discover the informal system, which is likely to be very different from the formal system

- Target specific steps for improvement

Process: A collection of interrelated actions, activities, steps, or tasks executed to achieve a specific result per customer requirement.

A process map is a visual representation of the steps required to complete a task. It displays the events sequentially using an agreed-upon set of symbols and, therefore, can be an effective analysis and communication tool. A good process map should allow people unfamiliar with the process to understand the interaction of causes during the workflow.

A process map may include tasks, activities, and/or decisions. It is therefore important to agree on a standard set of symbols and how they are used. Bill gave the handout shown in Figure 7.1 to his team to help them create a good process map.

Process mapping can be used in a variety of activities, including planning, analysis, writing procedures, and training operators.

Processes that are mapped are:
- Repeatable
- Predictable
- Improvable

Description	Symbol	Example
Labeled **terminal symbols** indicate the start and end points of the map.	START END	
Activity symbols depict single steps in the process and should follow a verb-noun format to describe the action being carried out. This format makes storytelling much easier.	Verb noun	Boil water
Decision symbols show when one or more alternative paths are available. Write as a question that can only have a close-ended answer. Use verb-noun nomenclature if possible.	Decision ?	Add sugar?
Link symbols indicate the map is too large for the width or height and a break must be inserted.	●	A Page 1 A Page 2
Flow lines show the connection and the direction of flow between steps.	 →	

Figure 7.1   Process mapping symbols.

Bill provided the following guidelines to his team:

- Use a pen and paper; do not use a computer. Quick and crude are better than slow and elegant. Get the ball rolling and keep it rolling.
- Use sticky notes to make adjustments and corrections.
- Use the conventional symbols consistently.
- Start with the people who actually do the tasks.
- Don't map based on information from written documentation; this will help you avoid mapping a theoretical instead of an actual process.
- Observe operators performing their regular duties in the regular work environment.
- Walk and record the real process.
- Observe and capture the sequence of computer transactions.
- Ask questions.
- Look for accountability systems, metrics, and checks and balances systems that may be in place.
- Pay attention to redundancies, particularly when looking at documentation and inspection steps that are repeated throughout the process.

- Note whether any employees show a lack of familiarity with documentation, procedures, processes, or local work instructions.

Bill coached his team as they developed the process map. He shared the following best practices, which help make process maps easier to understand for audiences:

- Sequence the boxes to be read from left to right and top to bottom

- Avoid having connection arrows intersect (Figure 7.2)

- Color code the connection arrows from decision boxes by using red for "no" and green for "yes" (Figure 7.3)

- Label all process steps using the verb-noun format

- Number the process steps on the map to make it easier for people to refer to them; this helps reduce confusion in case several boxes have the same labels (Figure 7.4)

- Use colors and/or swim lanes to highlight handoffs from one process operator to the next (Figure 7.5)

He also pointed out that no system works the way we (management) think it does and that the informal (de facto) system is likely to be very different from the formal (designed) system. In other words, every process map has at least three versions (Figure 7.6).

The team had a difficult time agreeing on several process-related details. Some claimed inaccuracies existed, while others admitted they were doubting their own knowledge of the process.

Bill had the team test the accuracy and completeness of the map by having an individual, Stephen, try to obtain the expected results by following the depicted steps. Stephen attempted to execute the steps exactly as they were documented while the rest of the team observed. Within a couple of steps the team realized that they had missed many key decision points and started talking over each other.

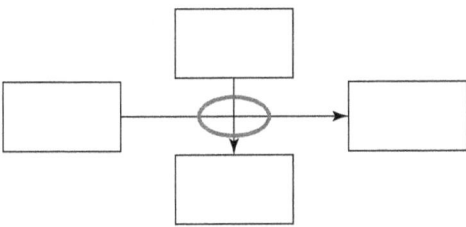

Figure 7.2 Intersecting lines.

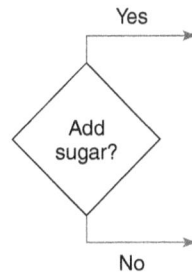

Figure 7.3 Yes/no decision box.

Figure 7.4 Numbered map steps.

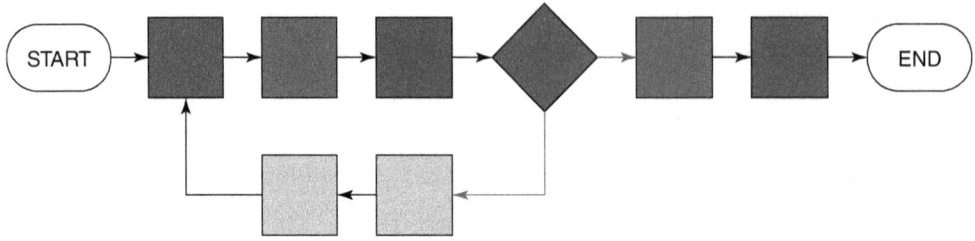

Figure 7.5 Swim lanes.

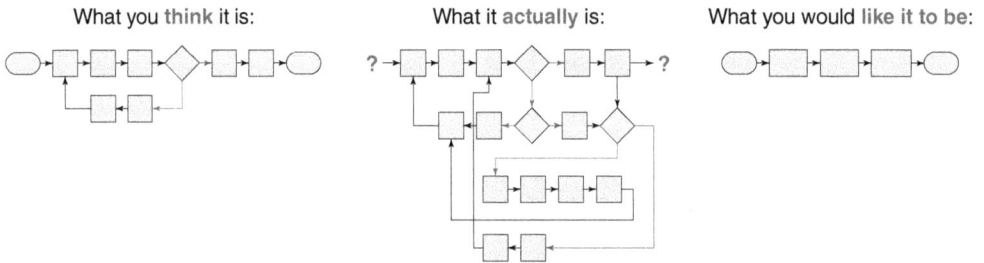

Figure 7.6 Process map versions.

Failure to obtain the expected result when following the steps of a process map could indicate any of the following:

- The process has variations (depending on who is performing the tasks)
- The map was drawn on a day when the normal process was not followed
- The map is incomplete or not properly drawn

Large teams sometimes have difficulty reaching consensus on the steps that constitute a process. Bill defused the situation by introducing to the team a tool called the seven ways.

8

Seven Ways

Seven ways: A group participation technique where participants are asked to re-create a process map seven times using sticky notes to represent the steps of the process. A different color of sticky note is used for each iteration; in other words, all the sticky notes from one iteration are the same color.

After each iteration, the facilitator collects the sticky notes and places them on a grid that captures the step numbers and that the participants cannot see as they work on their next iteration.

By the seventh iteration, the team generally feels it has captured all possible scenarios and each participant likely feels they have had the opportunity to express their views and are now more open to collaboration.

Bill had a large group of participants and decided to modify how his team would execute the seven ways. He divided the team into three groups, and each group was asked to produce two iterations of the process being analyzed. He distributed a different color of sticky note for each iteration to the groups.

Bill placed a seven ways grid on the wall. Each group added its sticky notes and presented its versions of the process map to the rest of the participants (Figures 8.1 and 8.2).

With these six iterations visible to the whole team, they were able to combine their thoughts to create one final consensus map as the seventh iteration (Figure 8.3).

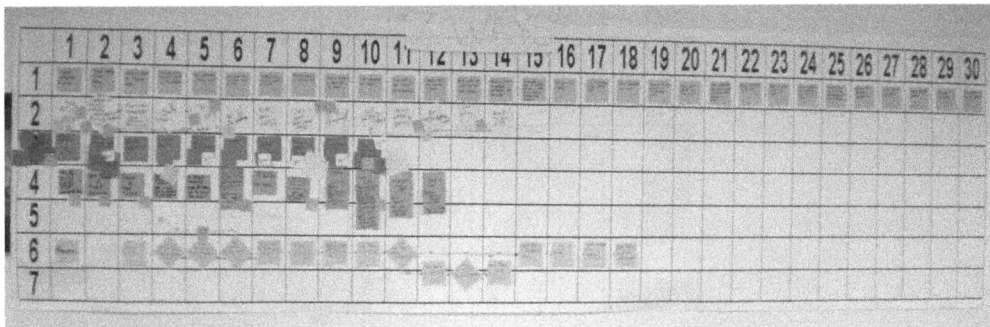

Figure 8.1 Photo of seven ways tool.

	1	2	3	4	5	6	7	8	9	10	11	12	13	14	15	16	17	18	19	20	21	22	23	24	25	26	27	28	29	30
Iteration 1																														
Iteration 2																														
Iteration 3																														
Iteration 4																														
Iteration 5																														
Iteration 6																														
Consensus																														

Figure 8.2   Grid for seven ways tool.

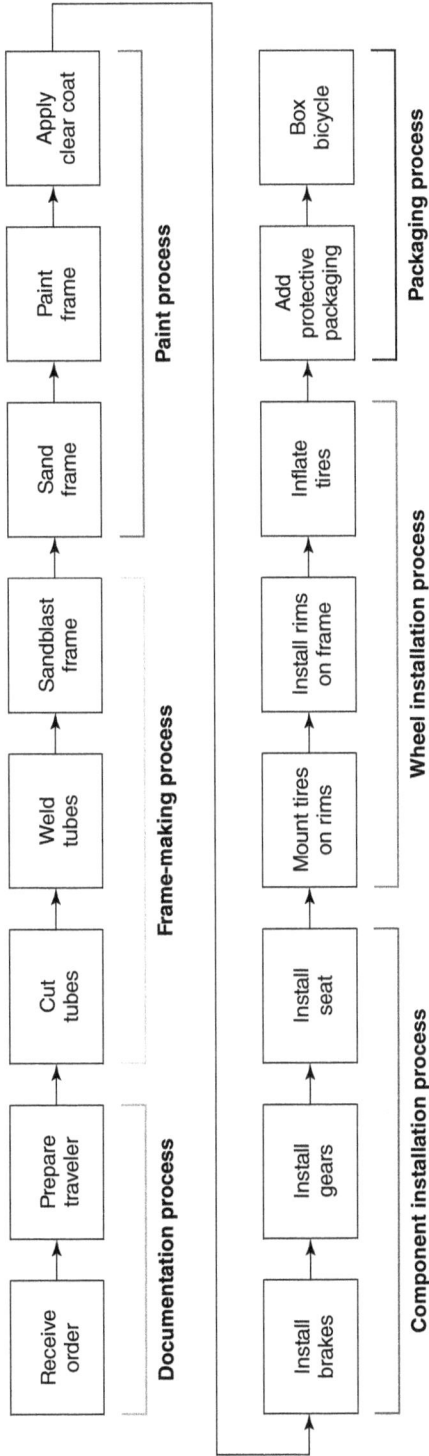

Figure 8.3 Velocycle consensus process map.

9

Detailed Process Map Analysis

Since the top-level process map is usually not very specific, the groups will need to drill down and gather more specific details to better understand the process.

In the interest of saving time and to leverage the expertise of each of the SMEs on his team, Bill divided the team into three groups and assigned a subprocess to each group for them to develop a detailed process map. All of these detailed process maps were later combined to obtain the map in Figure 9.1.

HOW TO USE A PROCESS MAP TO EVALUATE A PROCESS

Bill asked the team to leverage the process map to find opportunities for improvement by identifying steps that can be eliminated, rearranged, performed in parallel, simplified, expedited, or converted to less expensive operations.

In general, action verbs with the prefix "re" indicate some type of repetition because something wasn't done right the first time. These steps cause a backward loop in the process and are considered low-hanging fruit. Addressing them typically yields a large ROI in a short period of time. Figure 9.2 lists a few examples.

The team pasted numbered starbursts directly on the process map to tag process steps with opportunities for improvement (Figures 9.3 and 9.4).

> **Starburst:** A symbol in the shape of an explosion that is used by teams to visually indicate an opportunity for improvement at a step. They are usually numbered for easy reference.

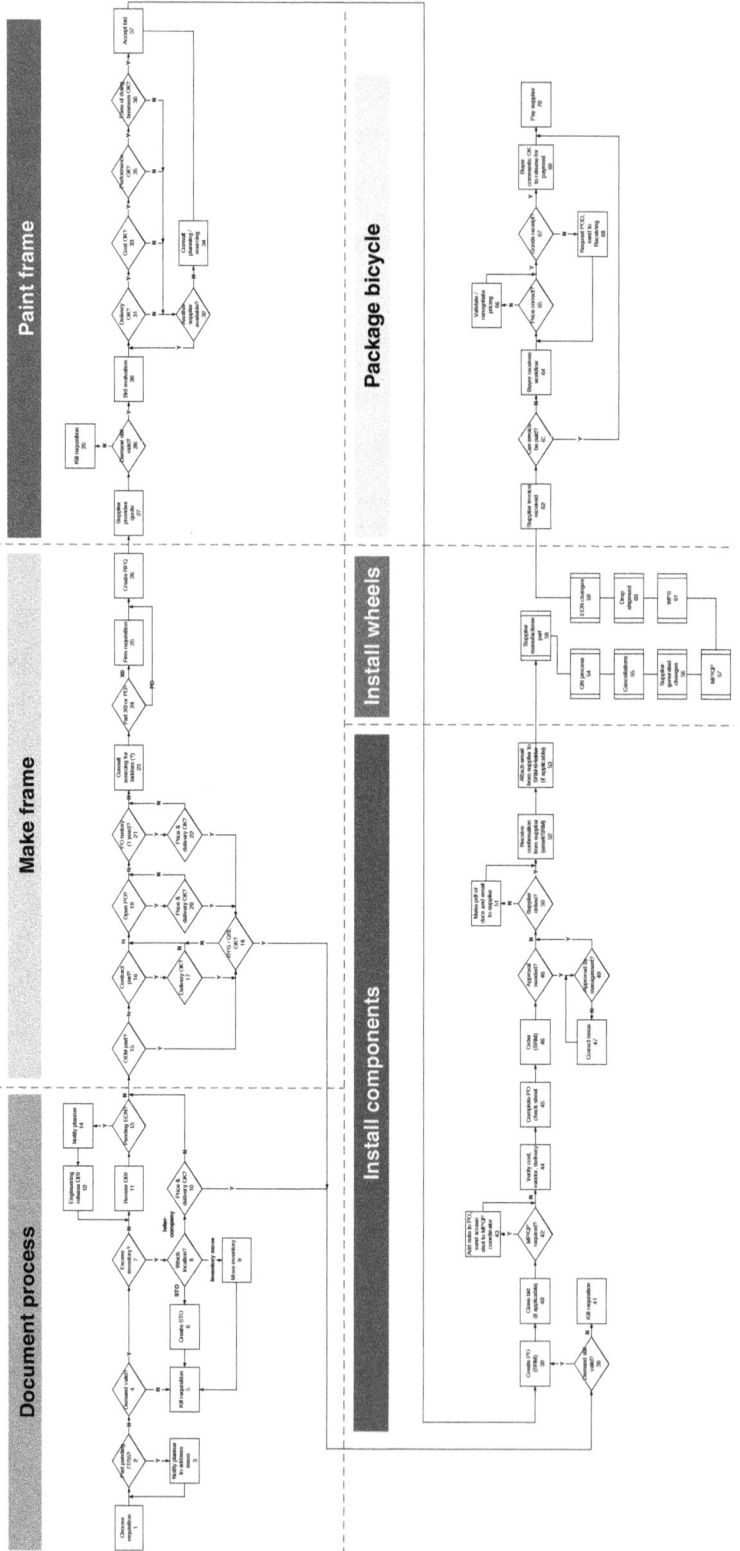

Figure 9.1 Detailed process map.

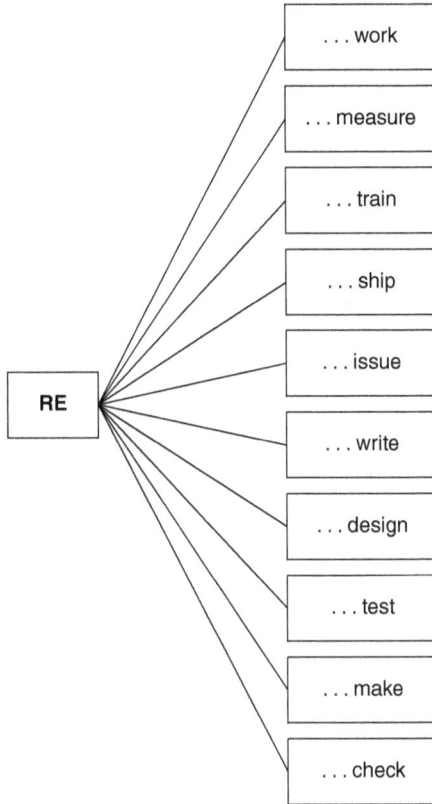

	... work
RE	... measure
	... train
	... ship
	... issue
	... write
	... design
	... test
	... make
	... check

Figure 9.2   Words with "re" as prefix.

Opportunities for improvement			
Starburst #	Step #	Step	Opportunity
1	21	Make frame	Recut tube
2	30	Paint frame	Repaint frame
3	39	Install components	Readjust brakes

Figure 9.3   Numbered opportunities for improvement.

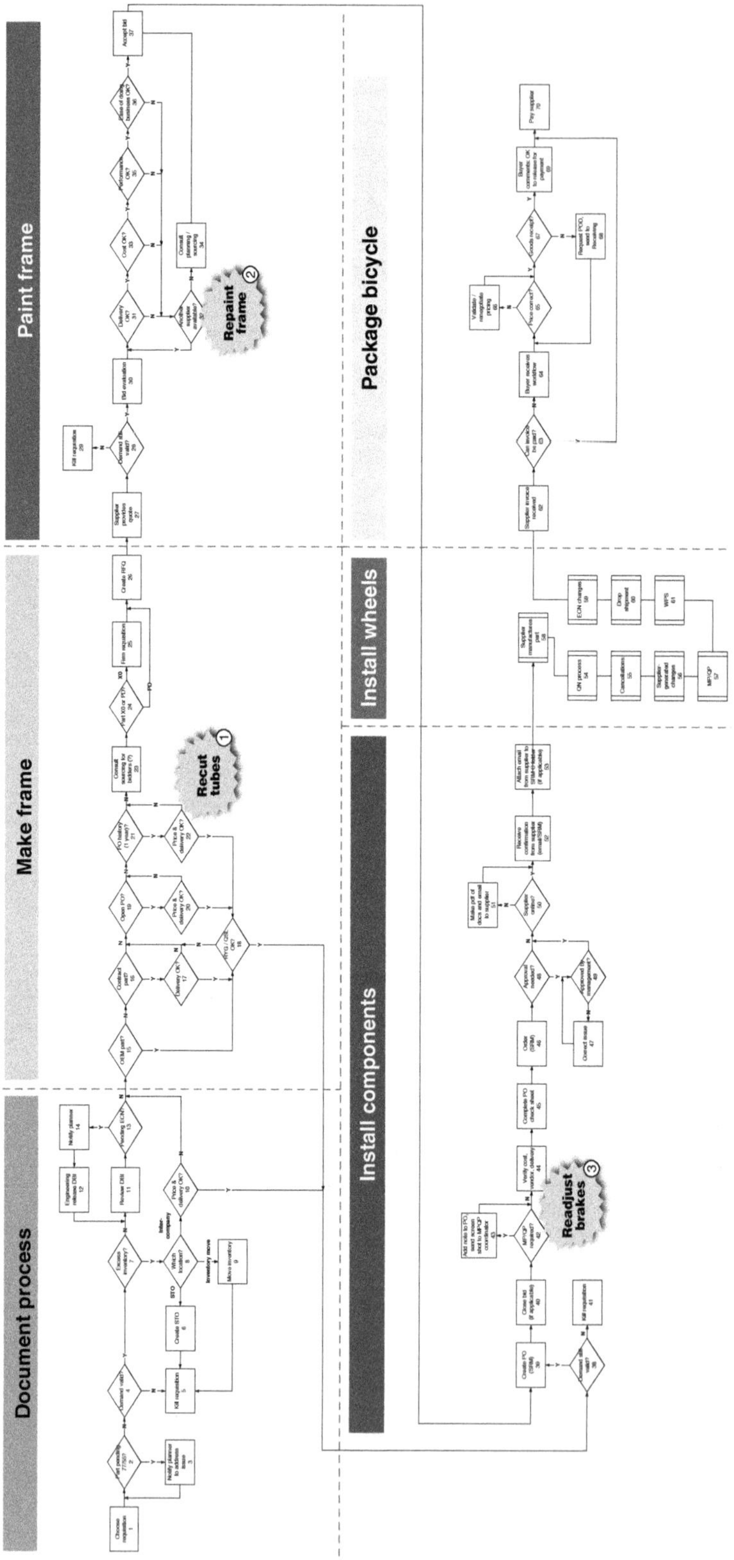

Figure 9.4 Detailed process map.

10
Waste

Bill knew there were more opportunities for improvement, but these opportunities were not visible to his team because of their lack of experience with lean. Therefore, he organized an impromptu training session to teach the team about waste and how lean aims to eliminate waste.

Waste: Any activity that consumes resources but creates no value for the customer.

THE SEVEN TYPES OF WASTE

Bill told the team about Taiichi Ohno, who classified waste into seven categories so that it could be easily identified and eliminated. He shared a simple mnemonic for remembering the seven types of waste and explained each type:

"Decreasing Overall Waste Is Up To Me"

   D = Defects

   O = Overproduction

   W = Waiting

   I = Inventory

   U = Unnecessary processing

   T = Transportation

   M = Motion

Defects

Defect: The waste associated with producing anything that does not meet customer requirements in terms of cost, timing, or any quality specifications.

Any activities to correct the defect are unnecessary processing and will add cost and delays. Rework robs resources and chokes flow. It must be minimized or eliminated.

Production example: scrap that has to be repaired or discarded

Transactional example: incorrect data entry, typographical errors, and illegible notes

Overproduction

Overproduction: The waste created by using resources to assemble unneeded parts ahead of schedule while production is waiting for other needed parts—in other words, producing more than what the customer wants.

Overproduction is highly wasteful to manufacturing plants because it prohibits the smooth flow of materials and degrades quality and productivity while consuming large amounts of cash.

Overproduction is considered the worst type of waste because its effects are the compilation of the effects of all other types of waste, and it requires additional materials, manpower, transport, and storage.

Production example: any non-optimally performed work, large deposits of material at each operation, large delivery quantities instead of frequent deliveries (in or out)

Transactional example: working in documents ahead of schedule while delinquent documents wait, working in the wrong documents at the wrong time, and generating more information, such as e-mail and reports, than what can be processed

Waiting

Waiting: The waste created by keeping a resource inactive because something required to perform the task at hand is missing.

Waiting results when a machine or person cannot perform a downstream activity because an upstream process is not delivering on time.

Production example: waiting for parts, labor, a spot in the shop to open up, tooling, or disposition

Transactional example: waiting for approvals, for blueprints, or for the copier to warm up

Excess Inventory

Excess inventory: The waste created by any inventory not required for current customer demand.

Keeping inventories of raw materials, work in process, and/or finished goods that are not needed ties up cash, requires space, consumes valuable resources not immediately needed, consumes productive floor space, delays problem identification and operating performance improvements, increases lead times, and inhibits communication.

Production example: accumulation of work-in-process inventory at individual operations; excessive amounts of finished goods inventory on shelves, racks, and floors

Transactional example: piles of unprocessed paperwork, large amount of unread e-mails, work backlog, pen inventory on desk

Overprocessing

Overprocessing: The waste created by unnecessary operations and the duplication of some tasks, such as inspection.

Overprocessing can be attributed to a lack of standardization or not enough education or training.

Production example: sorting parts that don't need sorting or individually wrapping parts that don't need to be wrapped

Transactional example: having 10 people check an internal e-mail for typos

Transportation and Conveyance Waste

Transportation waste: The waste created when transporting parts or documents from one operation to the next or to and from storage locations.

Excessive transportation and handling causes damage, consumes effort, impacts takt time, causes safety concerns, and may cause quality to deteriorate.

Transportation waste is countered by reorganizing operations according to sequence to create flow and minimize or eliminate the travel distance.

Production example: transporting products to an interim storage instead of directly from one process step to the next

Transactional example: mailing documents using the postal service instead of sending an e-mail or a fax

Motion

Unnecessary motion: The waste created by making extra movements; it applies to the human element, not the machine element.

Unnecessary motion adds to production time, increases employee fatigue, and causes frustration.

Production example: employees searching for tools, parts, or instructions

Transactional example: employees searching for documents

Unused employee talent (often referred to as the eighth waste): The waste associated with the failure to effectively engage people in the process, resulting in the underutilization of their talents, skills, and knowledge. Any time the team fails to make the most of an employee's potential capability is a lost opportunity.

BENEFITS OF ELIMINATING WASTE

Bill urged the team to eliminate waste to:

- Improve customer satisfaction
- Decrease cycle times
- Improve throughput
- Promote flexibility
- Have less work in process
- Generate more profit

Removing waste will allow you to do more with less.

SOURCES OF WASTE

Bill explained that waste is everywhere; it is in front of us so often that we don't always see it. It will result when any of the issues shown in Figure 10.1 remain unaddressed.

WHY CONTINUE DOING THINGS WE KNOW CAUSE OR LEAD TO WASTE?

Once the team members learned about waste, they saw it everywhere and wondered why these wasteful habits continue. Bill explained that as people live with their processes every day, they can't always perceive the progressive onset of wasteful activities. As a result, we often don't see what is really going on around us in terms of waste and ineffectiveness.

He quoted Yogi Berra to emphasize that "you can observe a lot just by watching." And if you don't ask why, it will happen again and will be "deja vu all over again."

Unaddressed issue	Resulting waste
Poor layout	Transport + motion
Long setup times	Waiting + inventory
Incapable process	Defects + unnecessary processing + motion
Poor maintenance	Defects + unnecessary processing
Poor work methods	Defects + unnecessary processing + motion
Lack of training	Defects + inventory + unnecessary processing + motion
Poor supervisory skill	Defects + overproduction + inventory + unnecessary processing
Ineffective scheduling	Defects + overproduction + unnecessary processing + inventory
Inconsistent performance measure	Defects + overproduction + unnecessary processing
Excessive control	Defects + unnecessary processing + overproduction
No backup/cross-training	Defects + waiting
Unbalanced workload	Defects + waiting + overproduction + inventory
No decision rules	Defects + unnecessary processing + waiting + inventory
No visual control	Defects + inventory
Lack of workplace organization	Defects + motion + overproduction + waiting + transport + inventory + unnecessary processing
Poor supplier quality	Defects + waiting + transport + inventory + unnecessary processing

Figure 10.1   Waste that results from unaddressed issues.

> Procedures and methods often change over time, and the original purpose becomes clouded or no longer exists. So ask "Why do we do that?" and "Why do we do it that way?"

He clarified that in some cases we fail to challenge unfounded beliefs such as:

- Maybe the customer requires it and we can't ask for a change
- Maybe the customer thinks they want it because we told them we could do it
- Maybe that is the way we have always done it

11

Value-Added/Non-Value-Added Analysis

Bill concluded his lesson on waste by mentioning that most companies waste 70%–90% of their available resources. Even the best manufacturers waste at least 30%.

Continuous improvement never ends, he harped. And we must seek to see waste, eliminate waste, create more value, and look for additional opportunities.

He transitioned and explained that each step of a process can be categorized as either value-added or non-value-added based on its contribution to the overall value of what is being created:

- Value-added activity (VA)

  - Any activity that changes size, shape, form, fit, or function of material or information (for the first time) to meet customer demands and requirements

  - Any service the customer is willing to pay for, including inspection or storage

- Non-value-added activity (NVA)

  - Any activity that consumes time or resources and that your process does not require (waste)

  - Any activity that does not satisfy customer demands or requirements

  - Any activity that a customer is not willing to pay for because it does not add value

In Figure 11.1, VA activities are in green and NVA activities are in red.

Essential non-value-added (ENVA) activities are a subset of NVA activities the customer is not willing to pay for but are required to meet third-party expectations or requirements. For example, activities that ensure safety or are required by regulatory agencies such as the Food and Drug Administration, the International

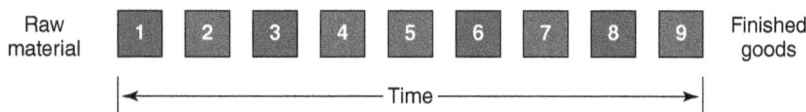

Figure 11.1 VA/NVA overview.

Organization for Standardization, and the American Petroleum Institute should be classified as ENVA.

> Any activity that does not add value in the eyes of the customer should be viewed as costly to the organization and counterproductive.

He printed and posted the diagram in Figure 11.2 to remind everyone of the following in the constant war against waste:

- VA activities should be optimized

- NVA activities result in defects, overproduction, inventory, and waiting that should be eliminated

- ENVA activities result in transportation, motion, and overprocessing waste that should be minimized

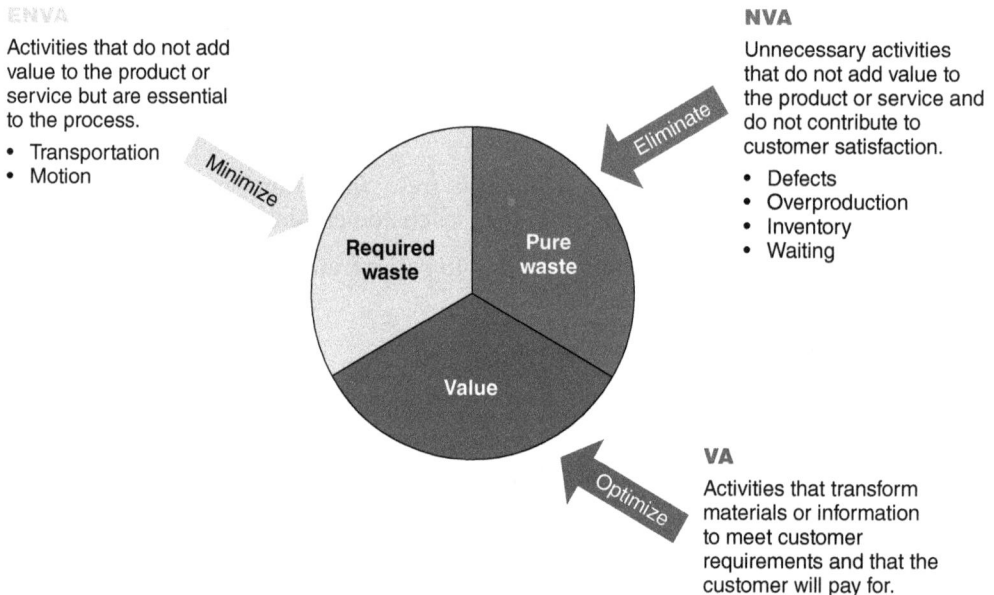

ENVA

Activities that do not add value to the product or service but are essential to the process.

- Transportation
- Motion

Minimize

Eliminate

NVA

Unnecessary activities that do not add value to the product or service and do not contribute to customer satisfaction.

- Defects
- Overproduction
- Inventory
- Waiting

Required waste

Pure waste

Value

Optimize

VA

Activities that transform materials or information to meet customer requirements and that the customer will pay for.

Figure 11.2   The constant war against waste.

Figure 11.3 depicts how long it takes to complete each of the activities. The color indicates whether it's a VA activity (green) or an NVA activity (yellow/red).

The figure illustrates how we can drastically reduce the amount of time required to complete all tasks by minimizing ENVA activities and eliminating NVA activities.

Bill warned the team about typical improvement initiative shortcomings that result from fixing parts of processes by implementing policy changes or adding reports or reviews without considering the impact on the entire process.

Many companies that don't perform a VA/NVA analysis try to improve the VA activities, which represent a small portion of the overall process, instead of trying to minimize or eliminate the NVA activities (Figure 11.4).

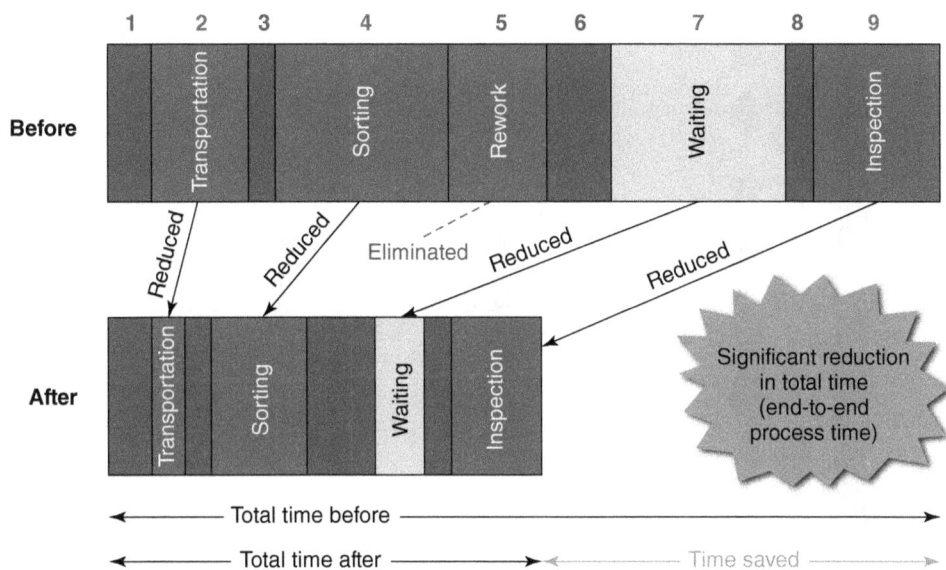

Projects that improve efficiency typically improve speed by reducing and eliminating waste.

Figure 11.3 Project efficiency gains.

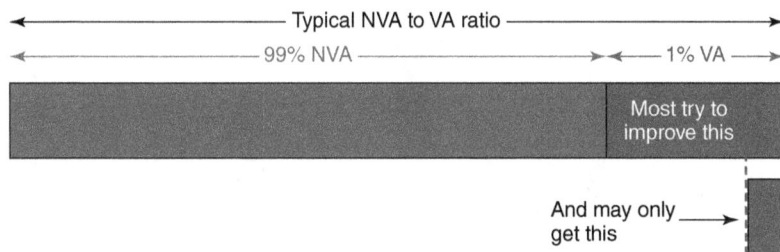

Figure 11.4 VA/NVA ratio.

Without looking at the entire process as an opportunity to do things differently, improvement efforts could result in:

- Suboptimization

- Spending in order to solve problems

- Processes that are not linked

- Failure to invoke systemic changes

- Taking special cause action on common cause problem

- Difficulties in sustaining

The team color coded the step numbers in red for NVA steps and in green for VA steps (Figure 11.5).

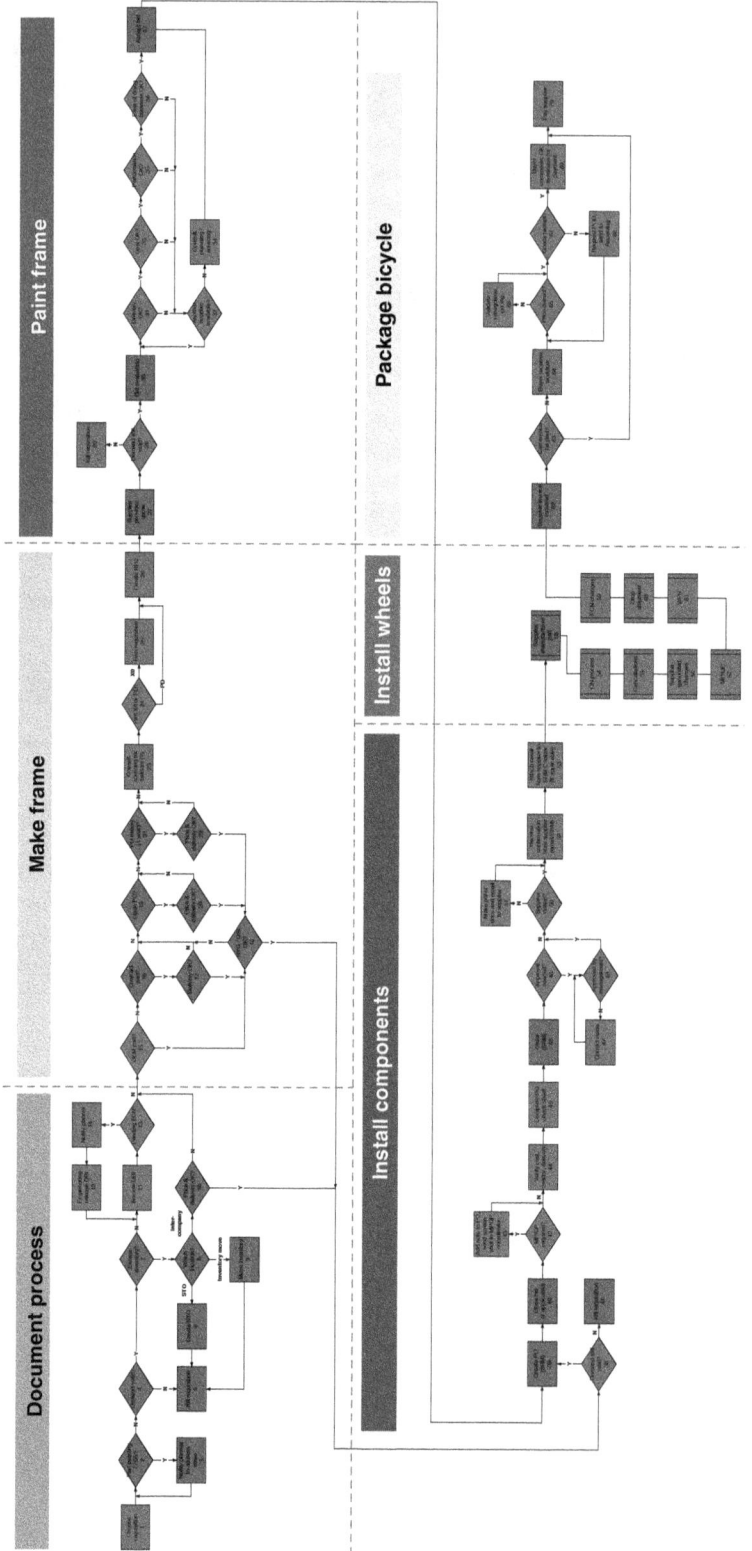

Figure 11.5   Map with VA/NVA analysis.

Table 11.1   VA/NVA results.

Velocycle assembly process		
Subprocess	**VA**	**NVA**
Documentation	9	5
Frame making	7	5
Painting	3	8
Component installation	4	12
Wheel installation	1	7
Packaging	1	8
Total	**25**	**45**

The team counted the number of steps for each subprocess and tabulated the results (Table 11.1). The team concluded that 45 out of the 70 actions performed to manufacture the product using the current processes are NVA. In other words, 64% of the actions have been identified as waste.

The first day of the event was full of excitement. The team members learned a lot about the processes being analyzed and had the following on the wall:

- The process scope

- Detailed process map

- VA/NVA analysis

The team was surprised the morning of day 2 when Bill explained that they still did not understand the process well enough to start making improvements. He used the drawing in Figure 11.6 as a road map to explain that every tool used has a ripple effect in the analysis and would be built on.

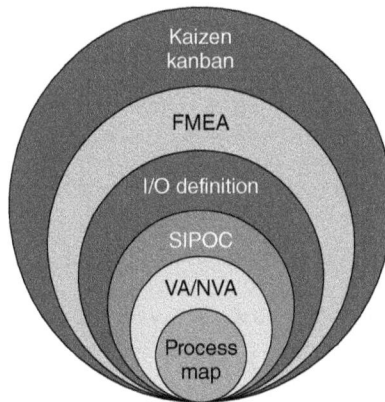

Figure 11.6   Tools.

12

SIPOC

Bill wanted the team to gain a better understanding of the connections between the process steps, the interaction between teams, and the impact of handoffs. Team members spent the next hour learning and applying a new tool called SIPOC, which is an acronym for supplier, input, process, output, customer (Figure 12.1).

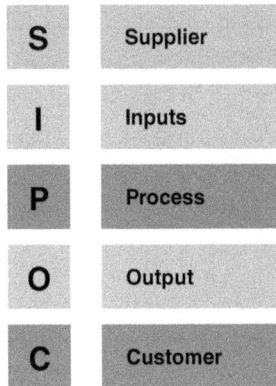

Figure 12.1 SIPOC.

The SIPOC creation process is more important than the diagram itself. Therefore, it is important to develop the SIPOC with the team.

Bill warned that the traditional SIPOC creation process tends to be focused on high-level process maps with fewer than 10 steps, and thus the information shown is often too general. To drive his point home, he guided the team in capturing the information in a traditional SIPOC format, as seen in Figure 12.2.

The team learned that a traditional SIPOC often fails to clearly show the linkages between suppliers, inputs, processes, outputs, and customers. Therefore, they developed an enhanced SIPOC methodology that enables capture of the information needed for a SIPOC visually and associates it with each corresponding process step. This approach builds on a process map to visually capture and display the suppliers, inputs, outputs, and customers associated with each step in a process and drives the team to ask the questions that need to be answered.

The tool that makes all this possible is the enhanced SIPOC box (ESB) (Figure 12.3). The following sections provide a step-by-step recap of how the team used the ESB and captured information one iteration at a time.

Suppliers	Inputs	Processes	Outputs	Customers
Rubbex	Boxes	Documentation	Wheels	Receiving department
MMA	Paints	Frame making	Frames	Welder
Pipex	Tubes	Painting		Shipping department
Sillas		Component installation		Paul
Pinturax		Wheel installation		Supervisor
		Packaging		

Figure 12.2 Traditional SIPOC for Velocycle.

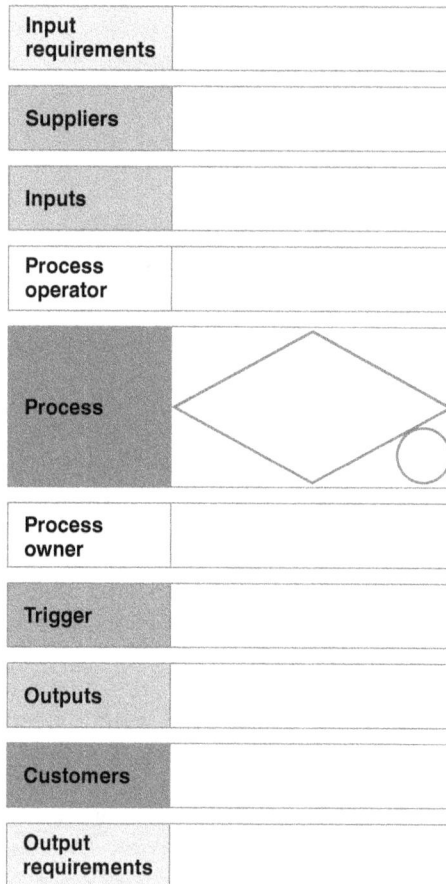

Input requirements	
Suppliers	
Inputs	
Process operator	
Process	
Process owner	
Trigger	
Outputs	
Customers	
Output requirements	

Figure 12.3 Enhanced SIPOC box.

ITERATION 1: ENTER PROCESS INFORMATION

The team started by taking the information for each process step that was captured on a sticky note and rewriting it on the blue section of the ESB labeled "Process" (Figure 12.4). They created one ESB to replace each of the sticky notes used to create the process map.

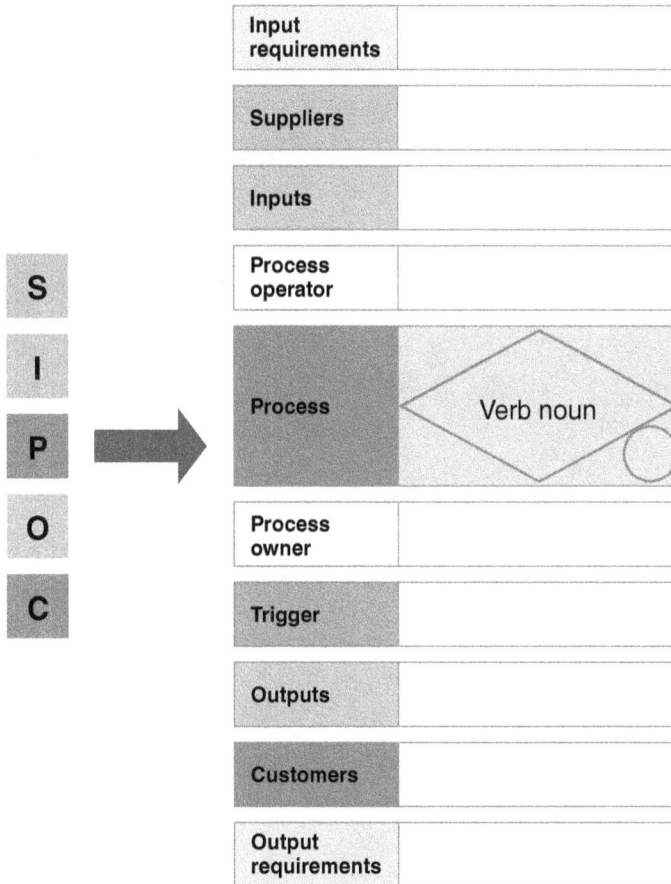

Figure 12.4   ESB—verb noun.

The VA/NVA analysis is represented by coloring the dot on the ESB red for NVA steps, green for VA steps, and yellow for ENVA steps.

This was also an opportunity for the team to double check that it had followed the verb-noun structure and to make adjustments as needed. Note that the "Process" section of the ESB can be used to show process steps and/or decisions.

The team organized the ESBs in sequential order and connected them using the appropriate connector protocol as with a regular process map (Figure 12.5).

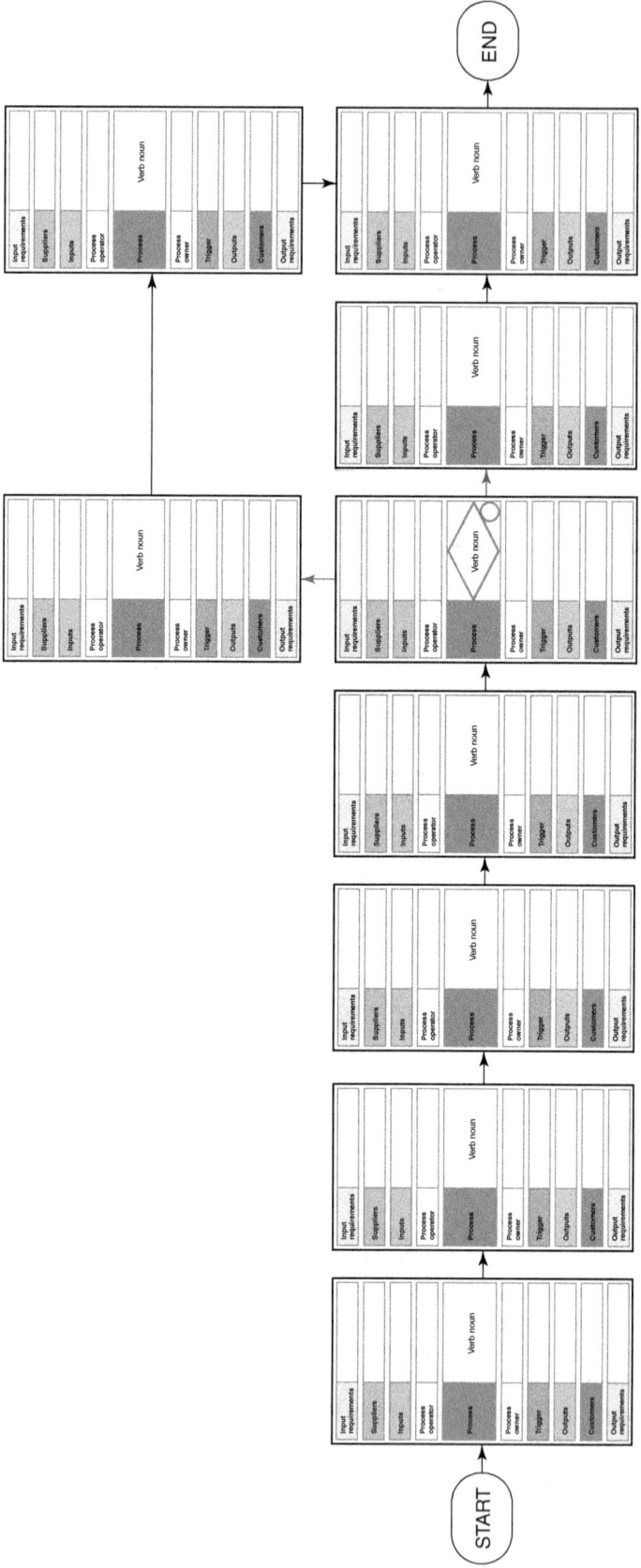

Figure 12.5 ESB applied to process map.

ITERATION 2: ENTER PROCESS OWNER AND PROCESS OPERATOR

In the second iteration, the team entered the process owners and process operators for each of the process steps (Figure 12.6). This step is a facilitation enhancement to a traditional SIPOC. Traditional SIPOCs don't require the process owners and operators to be added to each process.

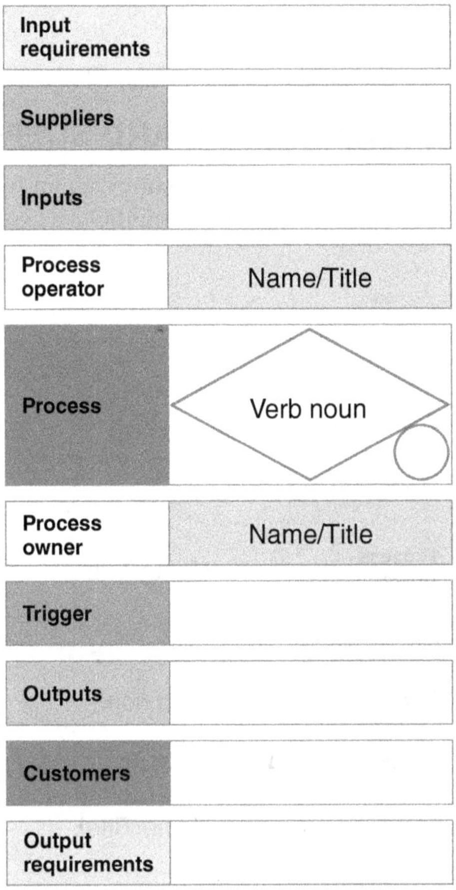

Input requirements	
Suppliers	
Inputs	
Process operator	Name/Title
Process	Verb noun
Process owner	Name/Title
Trigger	
Outputs	
Customers	
Output requirements	

Figure 12.6 ESB—process owner and process operator.

From a facilitation perspective, specifying the process operator and the process owner makes it easier for everybody to visualize who does each step and eliminates a lot of discussions when it comes time to identify suppliers and customers.

Process owner: The person who has the authority to make changes and is ultimately responsible for the performance of a process step as measured by key business indicators.

Process operator: The person who performs the actual work necessary to achieve the objectives of that process step.

ITERATION 3: ENTER THE OUTPUTS AND CUSTOMERS

In the next iteration, the team identified the outputs and customers for each step and added them to the appropriate section of the SIPOC template (Figure 12.7).

Outputs: Should be tangible and are usually written as nouns. They can be products, services, information, decisions, or documents.

Customer: Entity that receives the output of the process; they can be internal or external.

ITERATION 4: ENTER THE INPUTS AND SUPPLIERS

In the fourth iteration, the team identified suppliers and inputs for each step and added them to the appropriate section of the SIPOC template (Figure 12.8).

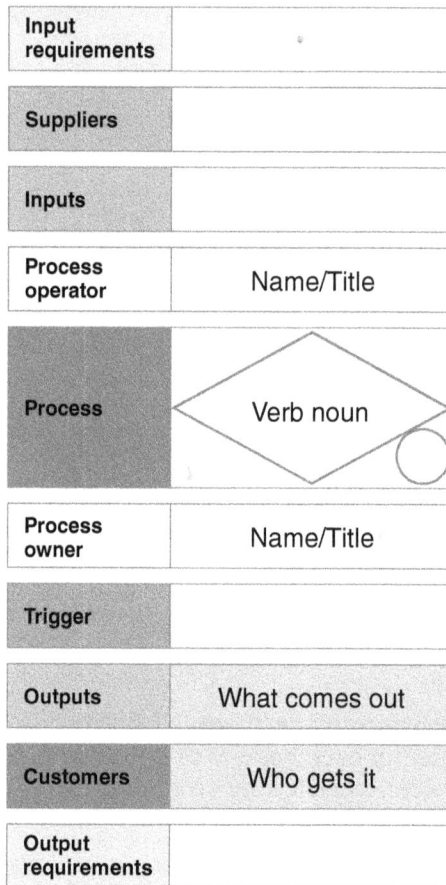

Figure 12.7 ESB—outputs and customers.

Input requirements	
Suppliers	Who gives it
Inputs	What goes in
Process operator	Name/Title
Process	Verb noun
Process owner	Name/Title
Trigger	
Outputs	What comes out
Customers	Who gets it
Output requirements	

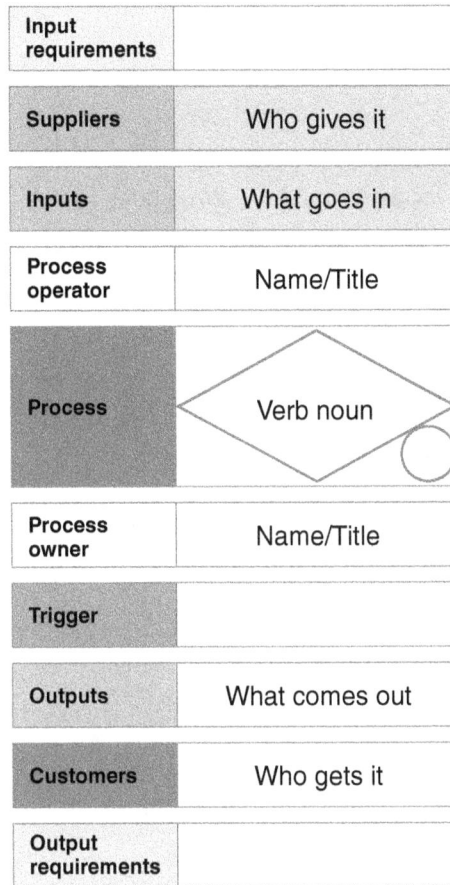

Figure 12.8   ESB—suppliers and inputs.

Inputs: Usually written as nouns and can be physical objects or information.

Supplier: Entity that provides input; they can be internal or external. In some cases, process operators are suppliers to themselves if they also carried out the previous operation.

Before moving on, the team ran through all the steps one more time to verify that the order of steps and the corresponding information were accurate.

ITERATION 5: ENTER THE TRIGGERS

The fifth iteration is an enhancement to traditional SIPOCs in which "triggers" are identified and added to each process step (Figure 12.9).

Trigger: Causes something to happen or lets the operator know the conditions are right to proceed to the next step.

Input requirements	
Suppliers	Who gives it
Inputs	What goes in
Process operator	Name/Title
Process	Verb noun
Process owner	Name/Title
Trigger	Why do we start
Outputs	What comes out
Customers	Who gets it
Output requirements	

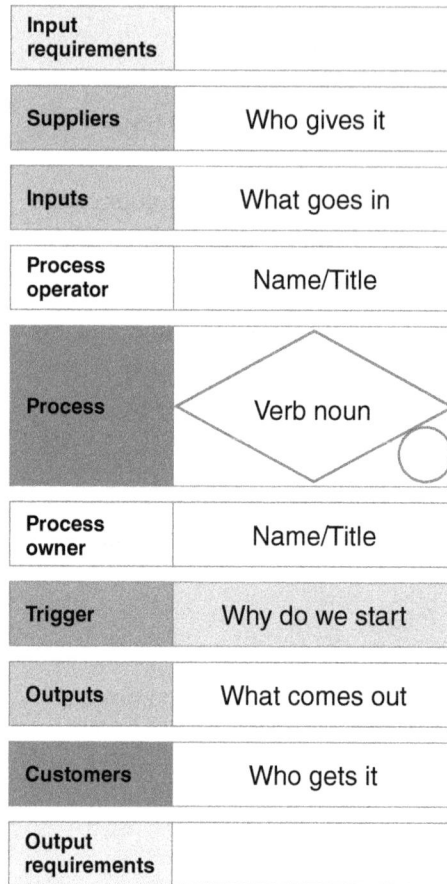

Figure 12.9   ESB—trigger.

Working backward from the last process step to the first, the team asked the process owners to identify the signal that lets them know everything is in place for them to execute their step. If a trigger exists but doesn't come from one of the previously identified steps, the map may be incomplete; a discussion should take place with the subject matter experts, and the necessary updates should be made to the map. Every time a trigger was found to be missing, the team added a starburst directly on the process map to indicate they had found an opportunity for improvement.

The absence of a trigger indicates potential delays in starting the next step, as the operator of that next step has no way of knowing the previous step was completed.

Note: Working backward helps you identify steps that you may have otherwise failed to notice.

ITERATION 6: CLEARLY DEFINE INPUT AND OUTPUT REQUIREMENTS

The sixth iteration is an enhancement to traditional SIPOCs that is essential in setting the team up for success when they create the failure modes and effects analysis in the steps to come.

The team reviews each of the inputs/outputs and adds the detailed specifications that must be met to ensure the material received by the operator is exactly what was expected (Figure 12.10). For example, if the input is water, clearly defined requirements could specify a volume, temperature, and/or purity. Specifying the volume of water can help reduce the likelihood of spills.

Input requirements	Specifications
Suppliers	Who gives it
Inputs	What goes in
Process operator	Name/Title
Process	Verb noun
Process owner	Name/Title
Trigger	Why do we start
Outputs	What comes out
Customers	Who gets it
Output requirements	Specifications

Figure 12.10 ESB—clearly defined specifications.

13
Handoff Map

While completing the SIPOC, the team members discovered that certain process operators seemed to be involved in several steps. Bill suspected that the people performing certain roles act as bottlenecks, as they are required to support the overall process and therefore must interfere frequently.

Bottleneck: A resource whose capacity is less than or equal to the demand placed on it and therefore limits the capacity of the overall process.

The team used the supplier/customer information from the SIPOC to create a handoff map (see Figure 13.1).

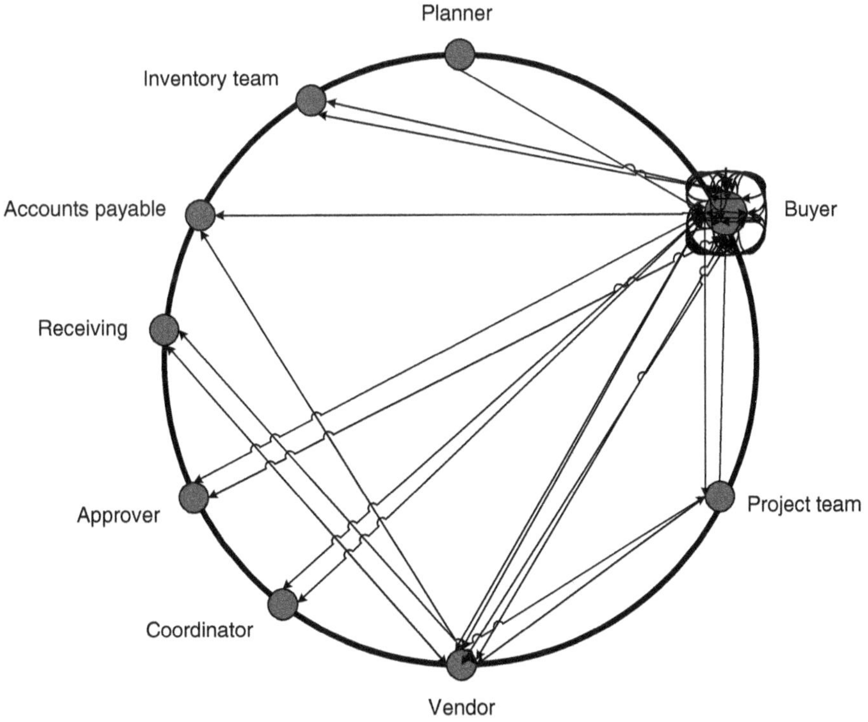

Figure 13.1   Handoff map

> **Handoff map:** A tool used to understand flow and eliminate unneeded hand-offs in order to minimize the number of NVA steps information must flow through.

Bill referenced the book *Lean Six Sigma: A Practitioner's Guide* (Wedgwood 2006) to provide step-by-step instructions:

- Draw a circle to represent the Velocycle internal universe

- Represent each of the functions involved in the process by a point on the circle, keeping them equidistant from each other, even if the distances vary in reality

- Trace each action as an arrow from one functional owner to the next

- If the interaction is within a group, or if one person does several tasks in a row, represent each task with an arrow to and from the same point

The handoff map revealed a high concentration of activities performed by the buyer, indicating frequent interactions between the buyer function and other roles. The results of the analysis of the handoff map are as follows:

- Of the 64 steps that are in scope, only 5 do not involve the buyer as either the supplier or the customer of that process. In other words, the buyer is involved in 92% of the actions.

- Of the 64 steps that are in scope, the buyer performs 45 in which he is the only direct customer. In other words, 70% of the actions done by the buyer are for the buyer.

- Of the 45 steps performed by the buyer for the buyer, 16 are VA. In other words, 64% of the actions the buyer does in which he is the only customer are not VA.

The team captured this information as a potential kaizen for the human resources department and placed a starburst directly on the process step on the map.

14
Failure Modes and Effects Analysis

After the team members clearly defined the input and output requirements for each process step, they had a much better understanding of the current state of the plant. They also had all the information they needed to analyze how the various steps could fail and the impact each potential failure could have on the overall process.

They used a failure modes and effects analysis (FMEA) to identify the potential risks associated with each step of the current process.

Failure modes and effects analysis (FMEA): Used to assess, manage, and reduce risk associated with the failure or potential failure of products, processes, services, and other systems before they occur. It may be used to define, identify, and eliminate known failures or errors in products before they reach the customer.

Risk: An intentional interaction with uncertainty that may result in the loss of something of value.

An FMEA is a proactive tool that enables the identification and prevention of process or product errors before they occur. The team used the FMEA to:

- Capture the collective knowledge of the team
- Identify process areas of concern using a logical and structured approach
- Improve the quality, reliability, and safety of the process
- Reduce process development time and costs
- Document and track risk reduction activities
- Help identify critical-to-quality (CTQ) characteristics
- Create historical records and establish a baseline
- Determine the highest-risk areas and prioritize actions to get the biggest bang for the buck
- Help increase customer satisfaction and safety

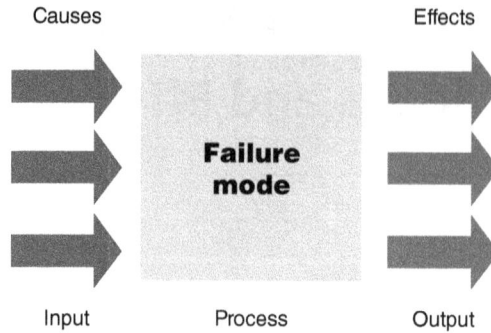

Figure 14.1   Failure mode.

Figure 14.1 illustrates how a failure occurs when one or more causes are present. Bill insisted that the team focus on the cause of the failure, not the effects. Failure may result from:

- Inherent product design flaws

- Manufacturing errors

- Improper customer usage and/or unplanned environmental use conditions

FMEA templates are widely available and contain fields to capture, at a minimum, the following information:

- Process step

- Potential failure modes

- Potential effects of the failure

- Frequency (F) of occurrence

- Severity (S) of the effects of the failure

- Detection (D) of occurrence

This information can be used to calculate the risk priority number (RPN).

FMEAs are traditionally completed in a group setting with one person capturing the information on a laptop or whiteboard while the other team members come up with each potential failure. This process is very time-consuming, places stress on the scribe, and leads to team member disconnect due to its slow speed.

The failure box (Fbox) was developed as an alternate approach that allows the entire team to be engaged and speeds up the process by dividing the team into subgroups and assigning each one a section of the overall process map (Figure 14.2).

The teams leveraged the clearly defined requirements captured in the ESB and built the FMEA directly on the previously created map using the following steps.

Failure mode		
Potential effect/consequence		
Frequency	Severity	Detection

Figure 14.2   Failure box (Fbox) template.

ITERATION 1: IDENTIFY ALL FAILURE MODES FOR ALL PROCESS STEPS

For each step on the map that was previously expended using an ESB, Bill asked questions such as:

- What could go wrong in this step?

- What could prevent this step from being completed?

- How could we fail to meet a requirement?

Team members wrote each of these answers in the failure mode area of a separate Fbox (Figure 14.3).

Each process step may have one or multiple failure modes, and you will need to fill out an Fbox for each of them.

Failure mode		
What could go wrong?		
Potential effect/consequence		
Frequency	Severity	Detection

Figure 14.3   Fbox with failure mode specified.

ITERATION 2: IDENTIFY EFFECTS FOR EACH FAILURE MODE

Each failure mode may result in one or more effects or consequences. An Fbox should be created for each consequence. For example, the failure mode "Part cut too short" has three potential effects/consequences: scrap, rework, and line stoppage (Figure 14.4).

Failure mode		
Part cut too short		
Potential effect/consequence		
Scrap		
Frequency	Severity	Detection

Failure mode		
Part cut too short		
Potential effect/consequence		
Rework		
Frequency	Severity	Detection

Failure mode		
Part cut too short		
Potential effect/consequence		
Line stoppage		
Frequency	Severity	Detection

Figure 14.4   Fboxes with same failure mode and three different effects.

Team members taped the Fboxes next to the corresponding ESB for each step on the map (Figure 14.5).

ITERATION 3: CALCULATE THE RISK PRIORITY NUMBER

Bill explained that an FMEA is used to determine which effect to address first based on the risk priority number (RPN) calculated for each step using the formula

$$RPN = Frequency \times Severity \times Detectability$$

The inputs to this formula are obtained by asking three questions for each process step, referencing a scoring guide, and assigning a numerical value as the answer to each question and placing it in the corresponding location on the Fboxes (Figure 14.6).

An FMEA can be performed for a product or a process.

Since the RPN is used for comparing alternatives with each other, it is important to ensure all participants are using the same scale to assign the values for frequency, severity, and detection. Bill distributed a reference document containing the three scales and explained that the white cells apply to a process and the gray cells apply to a product (Figure 14.7). We will discuss each of these three scales.

To assess frequency, the team referenced the frequency column to answer the question, How often does this consequence happen? (Figure 14.8).

Input requirements	Specifications
Suppliers	Who gives it
Inputs	What goes in
Process operator	Name/Title
Process	Verb noun
Process owner	Name/Title
Trigger	Why do we start
Outputs	What comes out
Customers	Who gets it
Output requirements	Specifications

Associated Fboxes:

Failure mode
Part cut too short
Potential effect/consequence
Scrap

Frequency	Severity	Detection

Failure mode
Part cut too short
Potential effect/consequence
Rework

Frequency	Severity	Detection

Failure mode
Part cut too short
Potential effect/consequence
Line stoppage

Frequency	Severity	Detection

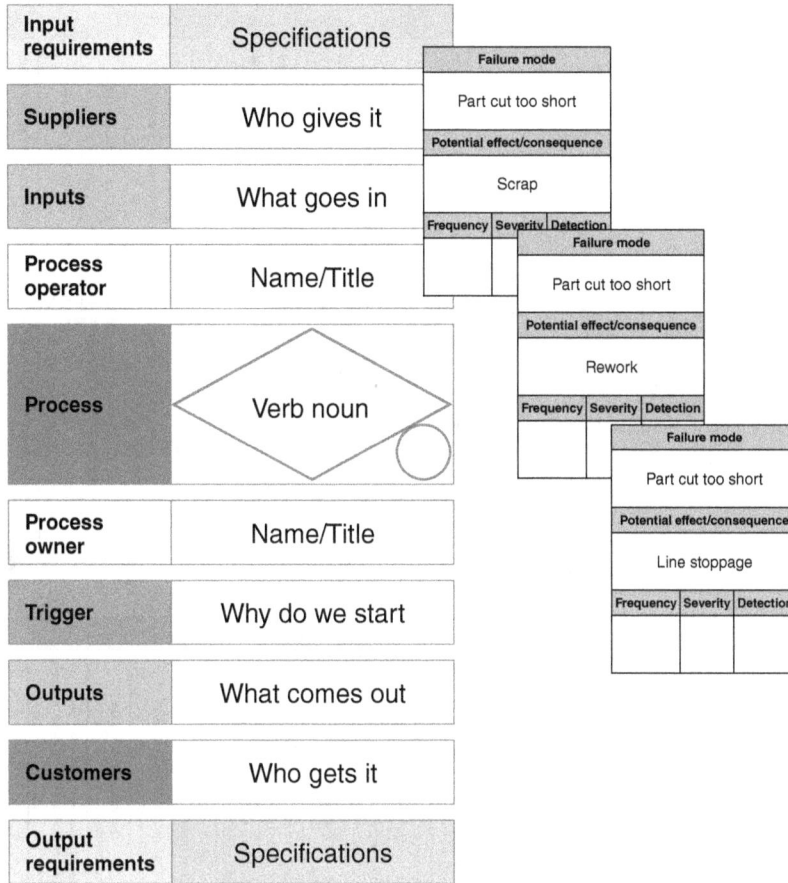

Figure 14.5 ESB with associated Fboxes.

Failure mode		
Potential effect/consequence		
Frequency	**Severity**	**Detection**
How often does it happen?	How bad is it if it happens?	Is it easy to know it happened?

Figure 14.6 Fbox with frequency, severity, and detection filled in.

Frequency

Probability that the potential failure will occur and thus result in the indicated failure mode

Category	Rank	Description
Very high: Failure is almost inevitable	10	Occurs very frequently (in the industry). 1 in 2 (fails 50% of the time).
	9	Occurs frequently. 1 in 4 (fails 25% of the time).
High: Repeated failures	8	Process has experienced much higher than normal failure rate. 1 in 10 (fails 10% of the time).
	7	Process has experienced higher than normal failure rate. 1 in 20 (fails 5% of the time).
Moderate: Occasional failures	6	Occasional failure. 1 in 100 (fails 1% of the time).
	5	Occasional failure. 1 in 1000 (1 failure per 1000 opportunities).
	4	Occasional failure, but not in major proportions. 1 in 5000 (1 failure per 5000 opportunities).
Low: Relatively few failures	3	Has occurred at least once in the industry. Minimum ranking for operator-dependent process is 3. 1 in 10,000 (1 failure per every 10,000 opportunities).
	2	Has occurred at least once in the industry. 1 in 100,000 (1 failure per 100,000 opportunities).
Remote: Failure is unlikely to happen	1	The team is not aware of this failure having ever occurred. Remote probability of occurrence; almost impossible. 1 in a million (1 failure per million opportunities).

Severity

Represents the seriousness of impact of the failure to the customer or to a subsequent process

Category	Rank	Description
Hazardous without warning	10	Affects process safety without warning. Will cause loss of life without warning (e.g., explosions or electrical shock will occur).
	9	Affects process safety with warning. Product is not usable, extensive damage.
Hazardous with warning	8	Loss of main functions; process is inoperable; safety issue. Product is not usable.
	7	Loss of main functions; safety concerns. Critical-to-quality product requirements not met.
Moderate	6	Failure causes customer concern or program impact; risk of nonlethal injury. Failure causes significant damage to product or equipment.
	5	Loss of noncritical function during processing; risk of nonlethal injury. Failure causes customer concern but will not cause a major failure of the end product.
	4	Loss of noncritical function during production or processing. Failure will not cause a major failure of the end product.
Minor	3	Minor effect on further processing or delivery to schedule. Minor effect on product performance. Loss of some noncritical functions.
	2	Very minor effect on further processing or delivery to schedule. Minor system errors that will not affect function or production.
No impact	1	Failure would have very little effect on further processing. Failure would have very little effect on product performance. May not even notice it happened.

Detection

The probability of detecting a defect (caused by the failure) *before* the item leaves the service or manufacturing location

Category	Rank	Description
Impossible to detect	10	Defect may elude even the most sophisticated detection technique. Very high probability of product leaving the manufacturing area containing the defect. Design control cannot detect potential cause/mechanism and subsequent failure mode.
	9	Detection may require special inspection techniques. High probability of product leaving the manufacturing area containing the defect. Very remote chance of detecting potential cause/mechanism and subsequent failure mode.
	8	Detection may require special inspection techniques. High probability of product leaving the manufacturing area containing the defect. Remote chance of detecting potential cause/mechanism and subsequent failure mode.
Very remote chance of detecting	7	Detection may require special inspection techniques. High probability of product leaving the manufacturing area containing the defect. Very low chance of detecting potential cause/mechanism and subsequent failure mode.
	6	The defect is somewhat more difficult to detect. Moderate probability of product leaving the manufacturing area containing the defect. Low chance of detecting potential cause/mechanism and subsequent failure mode.
Hard to detect	5	The defect is more difficult to detect. Moderate probability of product leaving the manufacturing area containing the defect. Moderate chance of detecting potential cause/mechanism and subsequent failure mode.
	4	The defect is somewhat more difficult to detect. Moderate probability of product leaving the manufacturing area containing the defect. Moderately high chance of detecting potential cause/mechanism and subsequent failure mode.
Easy to detect	3	Minimum ranking for operator-dependent inspection is 3. Low probability of product leaving the manufacturing area containing the defect. High chance of detecting potential cause/mechanism and subsequent failure mode.
	2	The defect is easily detectable; low probability of product leaving the manufacturing area containing the defect. Very high chance of detecting the potential cause/mechanism and subsequent failure mode.
Almost certain to detect	1	The defect is obvious, almost impossible that a defect would leave the manufacturing area. Almost certain of detecting potential cause/mechanism and subsequent failure mode.

Figure 14.7 FMEA calibration reference sheet.

Frequency		
Probability that the potential failure will occur and thus result in the indicated failure mode		
Very high: **Failure is almost inevitable**	Occurs very frequently (in the industry).	10
	1 in 2 (fails 50% of the time).	
High: **Repeated failures**	Occurs frequently.	9
	1 in 4 (fails 25% of the time).	
	Process has experienced much higher than normal failure rate.	8
	1 in 10 (fails 10% of the time).	
	Process has experienced higher than normal failure rate.	7
	1 in 20 (fails 5% of the time).	
Moderate: **Occasional failures**	Occasional failure.	6
	1 in 100 (fails 1% of the time).	
	Occasional failure.	5
	1 in 1000 (1 failure per 1000 opportunities).	
	Occasional failure, but not in major proportions.	4
	1 in 5000 (1 failure per 5000 opportunities).	
Low: **Relatively few failures**	Has occurred at least once in the industry. Minimum ranking for operator-dependent process is 3.	3
	1 in 10,000 (1 failure per every 10,000 opportunities).	
	Has occurred at least once in the industry.	2
	1 in 100,000 (1 failure per 100,000 opportunites).	
Remote: **Failure is unlikely to happen**	The team is not aware of this failure having ever occurred. Remote probability of occurrence; almost impossible.	1
	1 in a million (1 failure per million opportunities).	

Figure 14.8 Frequency calibration column.

To assess severity, the team referenced the severity column to answer the question, How severe is this consequence? (Figure 14.9).

To assess detectability, the team referenced the detection column to answer the question, How easy is it to detect this consequence? (Figure 14.10).

To make this activity even more interactive, Bill adapted a technique he had learned in an Agile project management course. He had each subgroup elect a scribe who would record the information directly in the Fbox. Bill also distributed special decks of cards with numbers from 1 to 10 to the other subgroup members for them to use to vote on frequency, severity, and detectability.

The rules he established were simple:

- The team members could not average the values on the cards. They had to decide on a specific number.

- The two people with the most extreme numbers would have an opportunity to explain their choices before the team voted again.

Severity		
Represents the seriousness of impact of the failure to the customer or to a subsequent process		
Hazardous without warning	Affects process safety without warning.	10
	Will cause loss of life without warning (e.g., explosions or electrical shock will occur).	
Hazardous with warning	Affects process safety with warning.	9
	Product is not usable, extensive damage.	
	Loss of main functions; process is inoperable; safety issue.	8
	Product is not usable.	
	Loss of main functions; safety concerns.	7
	Critical-to-quality product requirements not met.	
Moderate	Failure causes customer concern or program impact; risk of nonlethal injury.	6
	Failure causes significant damage to product or equipment.	
	Loss of noncritical function during processing; risk of nonlethal injury.	5
	Failure causes customer concern but will not cause a major failure of the end product.	
	Loss of noncritical function during production or processing.	4
	Failure will not cause a major failure of the end product.	
Minor	Minor effect on further processing or delivery to schedule.	3
	Minor effect on product performance. Loss of some <u>noncritical</u> functions.	
	Very minor effect on further processing or delivery to schedule.	2
	Minor system errors that will not affect function or production.	
No impact	Failure would have very little effect on further processing.	1
	Failure would have very little effect on product performance. May not even notice it happened.	

Figure 14.9   Severity calibration column.

Detection		
The probability of detecting a defect (caused by the failure) *before* the item leaves the service or manufacturing location		
Impossible to detect	Defect may elude even the most sophisticated detection technique. <u>Very high probability</u> of product leaving the manufacturing area containing the defect.	10
	Design control cannot detect potential cause/mechanism and subsequent failure mode.	
Very remote chance of detecting	Detection may require special inspection techniques. <u>High probability</u> of product leaving the manufacturing area containing the defect.	9
	Very remote chance of detecting potential cause/mechanism and subsequent failure mode.	
	Detection may require special inspection techniques. <u>High probability</u> of product leaving the manufacturing area containing the defect.	8
	Remote chance of detecting potential cause/mechanism and subsequent failure mode.	
	Detection may require special inspection techniques. <u>High probability</u> of product leaving the manufacturing area containing the defect.	7
	Very low chance of detecting potential cause/mechanism and subsequent failure mode.	
Hard to detect	The defect is somewhat more difficult to detect. <u>Moderate probability</u> of product leaving the manufacturing area containing the defect.	6
	Low chance of detecting potential cause/mechanism and subsequent failure mode.	
	The defect is somewhat more difficult to detect. Moderate probability of product leaving the manufacturing area containing the defect.	5
	Moderate chance of detecting potential cause/mechanism and subsequent failure mode.	
	The defect is somewhat more difficult to detect. Moderate probability of product leaving the manufacturing area containing the defect.	4
	Moderately high chance of detecting potential cause/mechanism and subsequent failure mode.	
Easy to detect	Minimum ranking for operator-dependent inspection is 3. Low probablity of product leaving the manufacturing area containing the defect.	3
	High chance of detecting potential cause/mechanism and subsequent failure mode.	
	The defect is easily detectable; low probablity of product leaving the manufacturing area containing the defect.	2
	Very high chance of detecting the potential cause/mechanism and subsequent failure mode.	
Almost certain to detect	The defect is obvious, almost impossible that a defect would leave the manufacturing area.	1
	Almost certain of detecting potential cause/mechanism and subsequent failure mode.	

Figure 14.10   Detection calibration column.

Once each effect was scored directly on the ESB on the wall, the subgroups were asked to transcribe their numbers to the Microsoft Excel template shown in Figure 14.11 for easy tabulation.

Failure mode	Potential cause(s)	Current process controls	Potential failure mode(s)	Potential effect(s) of failure	Frequency	Severity	Detectability	RPN

Figure 14.11 FMEA Excel template.

The spreadsheets from all the subgroups were compiled into a master spreadsheet and the resulting table was sorted by RPN to identify the highest-risk areas and where to focus actions first (Figure 14.12).

Sub-processes	Step #	Process step	Potential failure mode(s)	Potential effect(s) of failure	Frequency	Severity	Detectability	RPN
Frame making process	3	Cut tubes	Tubes too long	Rework and delay	3	2	1	6
Frame making process	5	Cut tubes	Tubes cut wrong angle	Scrap	5	4	5	100
Frame making process		Cut tubes	Tubes too short	Scrap	3	6	1	18

Figure 14.12 Compiled spreadsheets.

Referencing the Pareto principle, Bill explained that addressing the top 20% of the steps with the highest RPN could resolve 80% of the issues (Figure 14.13).

The team identified 200 failure modes for Velocycle; it decided to focus on the 40 failure modes with the highest RPNs.

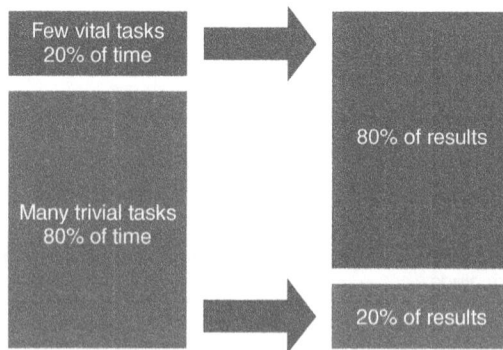

Few vital tasks 20% of time

80% of results

Many trivial tasks 80% of time

20% of results

Figure 14.13 Pareto's 80-20 rule.

15

Translating Opportunities for Improvement into Problem Statements

For each of the top 40 failure modes, Bill had the team members write a problem statement.

> **Problem statement:** A brief piece of writing used to inform the reader about the problem or issue, why it matters, and why a solution should be identified as quickly and directly as possible.
>
> A good problem statement answers the following questions:
>
> - What is wrong?
> - When did the problem happen?
> - Where is the problem occurring or where was it discovered?
> - How did the problem occur or how was it discovered?
> - What is the extent or magnitude of the problem?

Team members used the template shown in Table 15.1 to concisely capture their problem statements and document the link to the corresponding failure mode.

Table 15.1 Problem statement template.

Failure mode	Potential cause(s)	Current process controls	Problem statement

As the team members wrote their problem statements, some of them were more interested in solving the problems than they were in taking the time to clearly define them. They insisted on sharing their solution ideas. Bill, always the efficient facilitator, did not want to deviate from the agenda, but he also knew that ideas raised during the course of this discussion could be interesting and worthwhile to follow up on.

Bill had to find a way to quickly document these suggestions so they would not be forgotten. He also wanted to avoid sidetracking the entire team with an off-topic discussion and rebellions by team members who felt that their ideas were being ignored.

Table 15.2   Added potential idea column.

Failure mode	Potential cause(s)	Current process controls	Problem statement	Potential project idea

Bill solved the problem by adding a "potential project idea" column to the table that the team used as an idea parking lot (see Table 15.2).

> **Parking lot:** A large sheet of paper on the wall where off-topic issues are stored for future reference.
>
> The underlying purpose is to affirm people's value by acknowledging their ideas. Recording people's ideas tells them we want them to participate but don't have time for unrelated discussions. When people's ideas are validated and noted on the parking lot, they quickly acquiesce and refocus on the topic at hand. The parking lot brings everybody back on track.

Every time someone wanted to talk about solutions instead of writing a problem statement, Bill provided them with a sticky note and had them write down their idea. He placed these in the parking lot column.

16
Introduction to Kaizen

Team members were excited not only to have identified opportunities for improvement but also to have translated the most critical ones into potential project problem statements. They were all eager to start executing these projects one after another based on the RPN ranking. That's when Bill started speaking Japanese and initiated the team's education on the concepts of kaizen and kaikaku. Fortunately for the team, these were the only two Japanese words he used in his lesson.

He explained that the initial solutions, no matter how great, could always be further refined and improved over time and encouraged the team to adopt a mind-set of continuous improvement using a two-pronged approach:

- Kaikaku: for fast and significant improvements

- Kaizen: for slow and steady improvement

Bill explained that the Japanese word *kaikaku* means "radical change" and is used in reference to a business concept that aims to make fundamental and radical changes to a system or organization (Figure 16.1).

Kaikaku is a revolutionary approach to change, the sustainability of which is difficult to maintain without continuous improvements such as those accomplished through kaizens.

Kaizen is commonly used to describe a team approach to quickly break apart a process and rebuild it in order to function better (Figure 16.2).

Kaizen is a philosophy that advocates to continuously improve processes and address systemic and organizational opportunities for improvement by having employees at all levels of a company work together proactively to achieve regular, incremental improvements to the process. This creates an engine for improvement powered by the collective talents within the organization. Kaizen is evolutionary and focused on incremental improvements.

The Velocycle staff were determined to show how committed they were to embracing change and taking the company to the next level. From their perspective,

改革

Figure 16.1 Kaikaku.

改善

Figure 16.2 Kaizen.

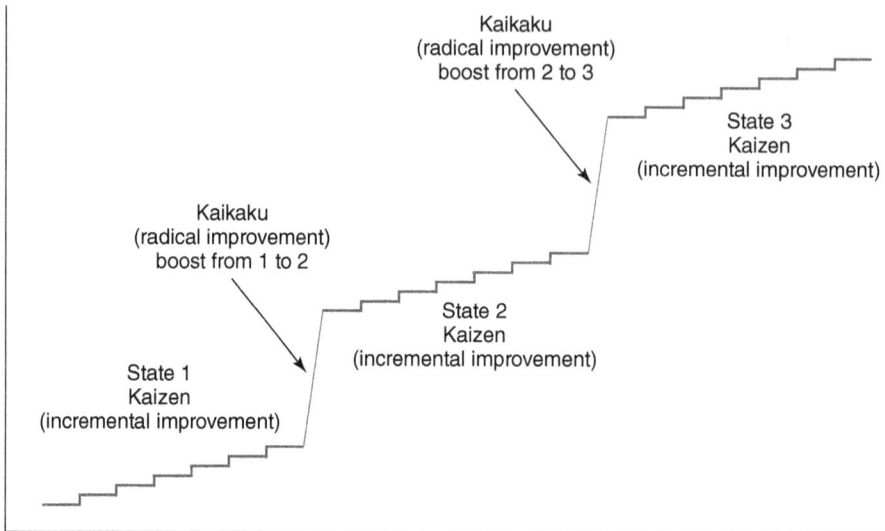

Figure 16.3   Kaikaku vs. kaizen.

only initiatives with a large impact on process performance would result in a significant return on investment they could boast to the management at SportsMax. In other words, they all expected to start with a radical change or kaikaku.

But this was not Bill's first rodeo. He drew the graph in Figure 16.3 to make his next point.

Before he started his explanation, he reminded everyone that the plant operators were key to the company's continued success and that they were predisposed to be wary of anything that could affect their livelihood—in particular, big changes because of the recent acquisition by SportsMax. Bill recommended a focus on earning quick wins to gain credibility and buy-in with small, easy, low-risk, and low-cost process improvements that can be easily evaluated with small and quick experiments. He promised that in due time the team would have the opportunity to complete a radical change.

Between the process map analysis, the VA/NVA analysis, SIPOC, and FMEA, the team members had identified a very large number of opportunities for improvements and prioritized the top 40 they wanted to address; however, they did not have unlimited time, money, or resources.

With the team members now understanding the difference between kaizen and kaikaku and the importance of the best sequence to follow, Bill proposed using a PICK chart to ensure they got the most benefits for their efforts (Figure 16.4).

PICK chart: A Lean-Six Sigma tool used to categorize process improvement ideas during a Lean-Six Sigma event. The purpose is to identify the most useful ideas. The 2×2 grid is normally drawn on a whiteboard or large flip chart. Ideas written on sticky notes by team members are then placed on the grid based on the payoff and difficulty level.

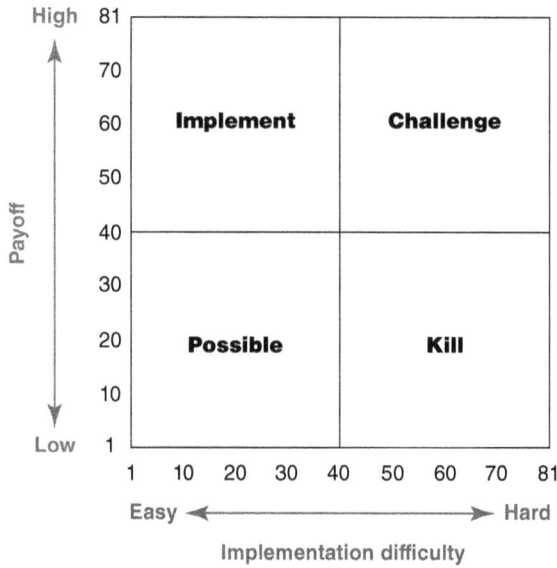

Figure 16.4   PICK chart.

The acronym PICK comes from the labels for each of the quadrants of the grid:

- Possible (easy, low payoff)
- Implement (easy, high payoff)
- Challenge (hard, high payoff)
- Kill (hard, low payoff)

In the interest of reducing the subjectivity sometimes associated with the use of a PICK chart, the team developed and used a more objective system to determine x (difficulty) and y (payoff) coordinates for each idea.

The implementation difficulty value (x coordinate) for any one idea is calculated using the formula:

$$\text{Difficulty} = \text{Duration Score} \times \text{Resource Type Score}$$

Table 16.1 shows implementation difficulty scores.

The y coordinate is the benefit value resulting from any one idea and is calculated using the formula:

$$\text{Benefit} = \text{Delay Score} \times \text{Impact Magnitude Score}$$

Table 16.1   Difficulty evaluation table.

Duration	Score	Resource type	Score
Short	1	Individual	1
Medium	5	Localized team	5
Long	9	Cross-functional team	9

Table 16.2   Benefit evaluation table.

Delay	Score		Impact magnitude	Score
Long term	1		Localized	1
Midterm	5		Department wide	5
Short term	9		Plant wide	9

Table 16.2 shows benefit value scores.

For example, one potential improvement project—develop a tool maintenance program—is as follows:

- Duration: Medium = 5

- Resource type: Cross-functional team = 9

  – Implementation difficulty = $5 \times 9 = 45$

- Delay: Midterm = 5

- Impact magnitude: Department wide = 5

  – Benefit = $5 \times 5 = 25$

Results are shown in Figure 16.5.

Referencing *The Lean Handbook*, Bill decided to organize the team's lean improvement needs based on the five levels of kaizen (Table 16.3). He provided each team member with the following information:

Lean improvement needs to be organized on five levels in most, if not all, organizations on a lean journey. The following sections discuss levels of improvement, in order of increasing scope and depth.

Individual (Point Kaizen)

At the individual workstation level, there are always opportunities to reduce waste—workplace organization, inventory and tool location, work sequence, ergonomics, poka-yoke, and on and on. The team leader has an important role to play here—encouraging, facilitating, and recognizing achievement and bringing individual improvements to the attention of others.

Work Teams (Mini Kaizen)

Work teams or groups that work in a cell or on a line segment undertake improvement projects affecting their collective work area. Examples include work flows, cell layout, line balancing, 5S, and quality improvements. These initiatives may be done on the fly or may be part of a one- to two-day mini kaizen.

Kaizen Blitz

A kaizen blitz is an event carried out in a local area, but it involves more time and outsiders. These events address more complex issues than what the work team can comfortably handle. For many companies, blitz teams are the prime

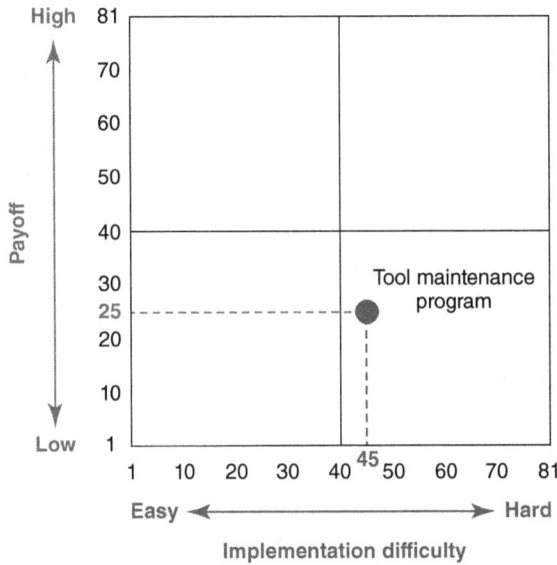

Figure 16.5 PICK chart example.

Table 16.3 Five types of kaizen.

Improvement level	Point kaizen	Mini kaizen	Kaizen blitz	Flow kaizen	Supply chain kaizen
Focus area	Individual workstation	Work cell or line	Complex local issue	Dock to dock value stream	Multiple companies
Typical tools and techniques used	Waste elimination Tool location Ergonomics Poka-yoke	Workflow Cell layout Line balancing 5S	Teams form for the specific purpose of the event	Cross-functional teams Process issues System issues Organizational issues	Part-time representatives from participating companies

engine for improvement. For this type of improvement, the team forms for the specific purpose of the event and disbands thereafter.

Flow Kaizen

Flow kaizen teams typically work across a full-value stream, taking weeks to months for a project. They are the prime engines for creating future states. Their targets are those set out in a future state and action plane activity. These teams are usually led by a project manager, often assisted by a champion, and sometimes mentored by consultants. The team comprises multidisciplinary and cross-functional members. Flow kaizen projects usually address process issues, system issues, and organizational issues.

Supply Chain Kaizen

Supply chain project teams comprise part-time representatives from participating companies within the value stream. They are focused on optimizing the entire value stream so that all within the supply chain can benefit from improvement. These teams usually have a project manager, typically from the original equipment manufacturer (OEM) company, and are supported by champions and consultants.

When you first start with improvements you may find it difficult to differentiate among short-, medium-, and long-term goals. It is best to just start and sort things out as you go. Here are a few guidelines to help you along the way:

- Manage each improvement with a single person
- Manage improvements visually in a single location
- Manage each improvement with a single schedule
- Hold review sessions weekly
- Avoid long projects (greater than 90 days)

Each organization needs to find its own style of improvement process and establish a standard method for everyone to follow.

Source: Manos and Vincent 2012, 235–236.

To help put things in perspective, Bill showed the team where each kaizen type typically falls on the PICK chart (Figure 16.6).

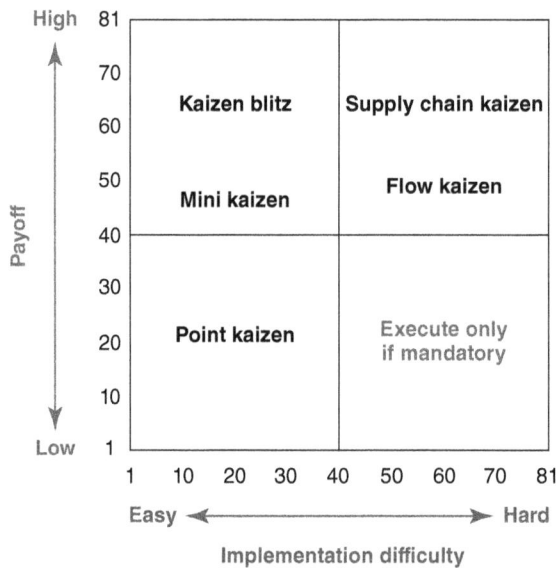

Figure 16.6 Kaizen and PICK chart.

17
Kaizen Kanban

Bill explained that as Velocycle progresses through its lean journey, there will be people who are at different stages in their lean journey; therefore, it is important to have each individual only execute projects they are capable of doing and that are important to the company.

To facilitate the selection and assignment process, Bill suggested the creation of a visual display of prioritized projects. A kaizen kanban, or improvement board, is a visual prioritized project pipeline and communication tool used by improvement teams to coordinate project selection and execution based on complexity and return on investment.

These boards should be visible to all levels of employees within the organization, and they follow the same principles used with traditional kanbans. The difference is that instead of telling the operators what to build next or what parts to retrieve, the boards tell improvement teams what pre-approved projects are most relevant to current business needs and are next in line for implementation.

In the spirit of continual improvement, the team did not want to waste time or money on creating complex displays. They opted to write each kaizen on a piece of 8½ × 11 card stock that had only the kaizen name and the problem statement (Figure 17.1).

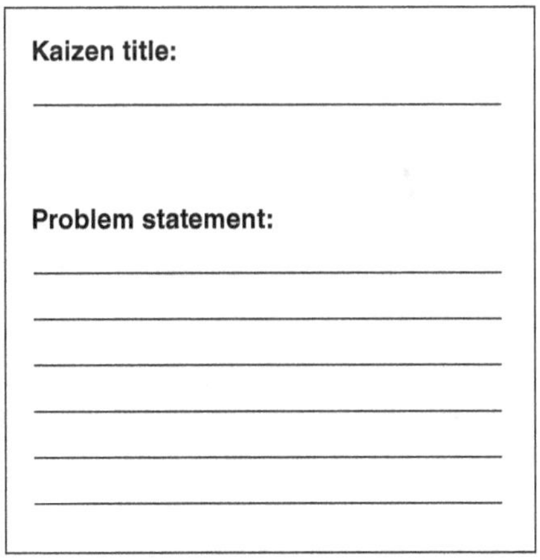

Figure 17.1 Kaizen card.

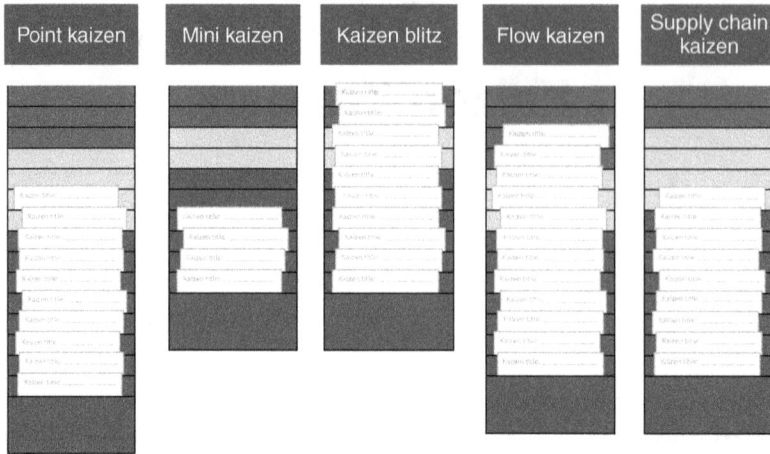

Figure 17.2   Kaizen kanban.

They used readily available wall folder holders to create the structure where the kaizen cards would be placed. To emphasize that there are five types of kaizen, they created five columns of file folders and labeled each one with a different kaizen type (Figure 17.2).

The team members surprised Bill by adding their own twist to the kaizen kanban board: they color coded the file folder holders to make it easier to see how the project relates to the three key business objectives corporate had set for the year:

- Reduce safety incidents by 20% (red)

- Reduce quality defects by 15% (yellow)

- Increase profitability by 17% (green)

18
Beyond the Kanban

Bill was proud of his team and of what they had accomplished together in such a short amount of time. Bill's assignment was to assess the situation and develop a list of the action items needed to turn the company around. When the team completed the kaizen kanban and presented it to Maggie Jean, who had flown to Miami to assess the progress personally, Bill's task was completed.

This was a very bittersweet moment for him, as he would not be part of the team executing the actual fixes. That task would be executed by his most successful mentee, Kiera Lamb, who was a senior principal engineer at SportsMax and an excellent project manager. She developed a unique project management–inspired process improvement methodology she called "Lean Agility." This methodology empowers self-managing teams to leverage lean, project management, and Six Sigma principles to successfully execute kaizens company-wide.

19
Step by Step

This book has provided a strategic approach for integrating fundamental quality tools such as process mapping, waste reduction, SIPOC, and FMEA into a Faster and Better methodology to create prioritized project pipelines by executing the following steps:

- Identify team

- Establish team rules and set operating mechanism

- Discuss the seven types of waste

- Identify first and last actions/operations and write them down on individual sticky notes

- Write down what happens step by step on a sticky note; be sure to use the verb-noun format

- Go observe the process or activity being performed

- Discuss VA/NVA

- Review each step and apply either a red sticker for an NVA step or a green sticker for a VA step

- Write VA/NVA report

- Write down who does each step (process operator) directly on the sticky note for the corresponding step

- Discuss the difference between process owner and process operator

- Write down on a sticky note who is the owner of that step and place the note under the process operator

- Introduce the concept of Supplier—Input—Process—Output—Customer—Trigger test

  - First iteration: Write down the outputs for each process step on a sticky note and place it under the corresponding step

  - Second iteration: Write down the customer for these outputs on a sticky note and place it under the corresponding step

  - Third iteration: Write down the required inputs for each step

- – Fourth iteration: Write down the supplier that provided each input

- – Moving backward from the end: Compare inputs for process N with the outputs from previous processes

- – Each input must be relatable to an output. If not, identify the source of the input. This may reveal whether a step was skipped. If a step was skipped, insert it now

- – Perform trigger test and identify gaps

- Discuss the handoff chart

- Create the handoff chart, identify the most frequent interactions, and identify bottlenecks

- Write clearly defined requirements for each input on a sticky note and place it under the corresponding input

- Review the requirements

- Discuss FMEA

  - – Develop severity, frequency, and detection scales adapted to the context of the area being analyzed

  - – For each SIPOC write all the different failure modes on individual sticky notes and place them directly on the process map

  - – Compile the information from the Fboxes in a Microsoft Excel spreadsheet and calculate the RPN for each failure mode

  - – Sort by RPN and select the top 20%

- Write a problem statement for each of the selected opportunities for improvement (OFIs)

- Rate potential project on difficulty and payoff

- Transfer to kaizen kanban board

- Present the kaizen kanban board to the stakeholders

- Select a team to carry out the kaizens

Part II
Facilitation

Everything we hear is an opinion, not a fact.
Everything we see is a perspective, not the truth.

—Marcus Aurelius

20

Facilitators Enable Doers to Do More

Throughout the case study in the first part of this book, we focused on addressing only one of the two deficiencies I had identified in the preface as a contributor to failed lean, Six Sigma, and continuous improvement initiatives: poor project selection.

In the second part of the book, the focus shifts to the lack of facilitation skills, which is the other major contributor to failed initiatives. In the chapters that follow, I share some key facilitation concepts that have contributed to my growth and my success.

Facilitation is a key skill, and those interested in catalyzing change must constantly develop and refine their mastery of that skill because lean, Six Sigma, and most continuous improvement tools are most effective when used in a group setting.

According to the website Vocabulary.com, the verb "facilitate" comes from the Latin *facilis*, for "easy." It means to make something easier or more likely to happen. By extension, business facilitation focuses on making meetings more productive. The facilitator is charged with performing all the tasks necessary to make the group's work more effective; this includes structuring and staging meetings for success and eliminating all sources of hassles, delays, and inconveniences. This individual ensures that systems are in place to allow each member to contribute and for each idea to be explored.

21
Facilitator Guidelines

During a meeting, good situational awareness is of utmost importance, and a good facilitator should:

- Be informed and capable of clearly explaining the purpose of the initiative and its benefits

- Remain neutral at all times and not seek to influence a decision one way or another

- Never do for the group what it can do for itself

- Be unobtrusive and not attempt to simultaneously be a participant

- Focus on the process and let the team worry about the content

- Set the tone and direction for the meeting

- Take the appropriate course of action to keep the meeting on track

- Anticipate and accept that not everybody will be on board with every new idea

A facilitator must understand human behavioral concepts, for they are facilitation building blocks and should be considered prerequisites to facilitation.

22
Paradigm Shifts

Each individual has an unspoken set of rules known as a paradigm, which sets boundaries while simultaneously providing problem-solving guidance. In some cases, it's just a heuristic or rule of thumb, and those who use them are usually open to being challenged given the self-evident vagueness of these beliefs. In other cases, they have become a mind-set, a habit, or a specific view of the world. These are harder to change, making it impossible for others to see what is so obvious to those individuals (Figure 22.1).

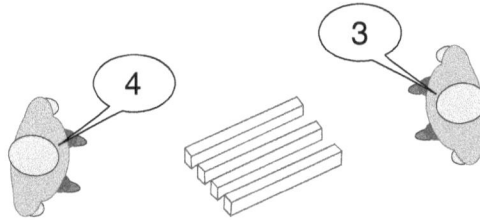

Figure 22.1   Paradigm.

When we are first exposed to a new idea, we are more likely to embrace and even adopt it if it aligns with what has worked for us in the past. On the other hand, ideas that require us to change the way we look at things create uncertainty and may require conscious effort if we are to accept them.

In his book *Paradigms: The Business of Discovering the Future*, Joel Barker (1993) explains a paradigm shift as an event causing everybody to go to zero, a new starting point, to illustrate how much courage is necessary for some practitioners to let go of the old way and accept a new reality.

New or revolutionary ideas challenge well-established truths at times. Therefore, when guiding a group of people attempting to cause or provoke a change, it is very important to understand the underlying beliefs that drive their decisions, if you're to accomplish long-term acceptance and buy-in.

23
Team Life Cycle

Without facilitation, there cannot be team building. The facilitator must keep the team moving forward by ensuring that interactions between members stay cordial, professional, and productive.

As the team works together and evolves, the interactions of the members, the way they carry themselves, and the way they behave also mature along a predictable pattern. The facilitator should be familiar with the concept of the team life cycle, which separates team evolution into five phases, known as Tuckman's ladder: forming, storming, norming, performing, and adjourning (Figure 23.1). The facilitator should know what behavior to expect from participants at each stage and what tools or techniques may have to be employed.

Performing
- Teamwork
- Cohesiveness
- In equilibrium
- Value adds

Norming

Adjourning
- Separation anxiety
- What next?
- Unease
- Dissatisfaction

Forming
- Excitement
- Anticipation
- Anxiety
- Optimum

Storming
- Shared goals
- Team cohesion
- Trade-offs
- Acceptance

- Reality sets in
- Communication gaps
- Adjustment anxiety
- Frustration, conflicts

New team formation

Existing co-located team converted to virtual team

Figure 23.1 Tuckman's ladder.

24

Setting the Stage for Successful Meetings

MEETING FACILITATION PREWORK

Just as runners training for a marathon start preparing weeks before toeing the start line, facilitators often complete prework activities weeks in advance to ensure the engagement of the participants and a successful meeting.

Some of the key preplanning activities they may perform include:

- Identifying a meeting time that accommodates the most participants (this is particularly important for virtual teams that may be in different time zones)

- Securing a physical location for a meeting that minimizes travel time and disruption to the most participants, or setting up a virtual meeting place

- Making sure all relevant information is available to participants prior to and during the meeting

- Preparing a meeting agenda

- Ensuring that support materials such as projectors, whiteboards, markers, sticky notes, and flip charts are available

- Obtaining refreshments for meetings that extend past mealtimes

SESSION GUIDELINES AND GROUND RULES

Setting rules with the team early on in the process provides the facilitator and members with a road map for handling crises, thus reducing the likelihood of conflict.

For short sessions, the facilitator can leverage a standard list of generally accepted meeting guidelines to set the tone and get the meeting moving. For longer sessions or workshops, the facilitator may encourage the team to create its own meeting ground rules.

Guidelines are generic in nature and promote the creation of a serene and peaceful atmosphere, which encourages creativity. Good guidelines take into account that people have responsibilities and needs beyond the scope of their involvement with a particular initiative.

Example of Session Guidelines

- Be kind and considerate

- Turn off computers and silence cell phones

- Excuse yourself and take urgent calls outside the room

- Enter into discussions enthusiastically

- Give freely of your experience

- Allow and encourage others to contribute

- Ask questions when you do not understand

- Appreciate the other person's point of view

- Provide constructive feedback—and receive it willingly

- Confine your discussions to the topic

- Keep things straightforward and simple

- Be innovative—encourage creative solutions

- Most of all, have fun!

Ground rules should be developed and agreed to at the beginning of the kickoff meeting; this helps set the tone and secure the buy-in of all team members. It is important to have a flip chart available to capture the ground rules as they are created.

This is also the perfect opportunity to emphasize the importance of:

- Ensuring confidentiality (in other words, "Our discussions remain in this room")

- Focusing on the task at hand ("If you had something more important to do, you would not be here right now")

- Getting a good return investment for our time and effort

PARTICIPANT INTRODUCTION

Getting to know the participants is very important. Understanding why they are here, what drives them, their preferred communication style, the role they play on the team, what they bring to the table, and how they define success for this task provides the facilitator with the insight needed to best align interests and avoid conflict. The facilitator may request that each participant share the following information about themselves with team:

Name: People who are sensitive about the pronunciation of their name can use this opportunity to let the team know how to address them.

Job title: This may help the team understand the perspective from which the individual views the world. A safety manager will look at the world from a different set of lenses than that of a quality manager.

How long you have been with the company: A longtime employee may bring historical information that could help the team avoid repeating mistakes of the past, while a brand-new employee may bring experience acquired at other workplaces that the team can use for benchmarking. Participants may also bring preconceptions that will have to be addressed at the appropriate time.

Your favorite inventor or invention (or favorite actor/sports team/movie): This question is used as an icebreaker.

Why you are here: This forces the person to share with the team their reason for being on the team, their "what's in it for me" (WIFM), and why they care about this project.

What would make this session a success for you: This may be one of the most important bits of information obtained from this exercise, as it lets the other team members know what the individual is working toward. Record these statements of success on a flip chart and use them to align and get buy-in from the team members.

ICEBREAKERS

An icebreaker is a mechanism by which the facilitator can help a participant feel comfortable enough to introduce themselves and share a fact that helps the team get to know them better.

Icebreakers are also an opportunity to have some fun and do some team building. The facilitator may organize a team activity such as solving a puzzle to get the participants to interact and work toward solving something that is not work related and intimidating.

Icebreaker Ideas

- Ask each team member to choose between two impossible situations, such as being stranded with lots of water and no land or being stranded with lots of land and no water

- Ask each team member to share some interesting trivia about themselves, such as:

  - Their favorite cartoon character, movie, or song

  - What musical instrument they wish they played

ADDITIONAL MEETING ROLES

Even the most seasoned facilitator cannot do everything on their own and must recruit the support of participants to make the session as effective as possible.

The facilitator should not be a doer while facilitating; the facilitator helps doers do more. Team members should fill the following roles:

- *Timer:* This person makes sure that the facilitator manages time properly and that the team uses only the allocated time for each point on the agenda.

- *Note taker/scribe:* This person captures meeting minutes that, once compiled, are shared with the team. Note that the facilitator should not be the note taker or scribe, since the facilitator should focus on the process. (For larger groups I usually request the assistance of two scribes to capture the team's ideas on flip charts.)

PARKING LOTS AND RABBIT HOLES

It is important for all participants to have a chance to have their say. In some cases, participants may completely shut down and stop contributing if they think they are not being given the chance to speak or feel they are not being listened to. Having their way isn't as important to them as having the opportunity to say what is on their mind and being heard.

In other cases, participants are just talking off topic. By using an idea parking lot and a rabbit hole interruption system, you can avoid losing both time and participants.

Creating an idea parking lot is as simple as placing a large piece of paper on the wall and writing "parking lot" on it. It is important to explain to the team that the parking lot will be used to capture ideas that are not relevant to the current discussion but may be useful in the future. Participants are encouraged to write down their ideas and place them in the parking lot, especially if their idea comes at an inopportune moment that could derail the group. Stress to the participants that these ideas will be revisited before the end of the session.

At the beginning of every meeting, I present each participant with a paper fan with an image of a rabbit on it, asking, "Do we need to go down this rabbit hole now?" (Figure 24.1).

I tell the participants that the fans should be raised whenever they feel the group has gotten off track in its discussion. The fan was not designed to question the validity of a point being made, just the timing. The first use of the fan often results in a lot of laughter, as it seems a bit silly and childish. But it helps keep the group on track and allows somebody to interrupt someone else without being rude.

Figure 24.1   Rabbit hole paper fan.

As soon as a rabbit is raised, the team is expected to stop its discussion immediately and vote to either continue the discussion because it's not out of place or move the idea to the parking lot.

FLIP CHARTS AND WHITEBOARDS

The most important benefit of using a flip chart, chalkboard, or dry erase board is that it results in the shared visualization of the opportunity at hand, thus eliminating the need for the attendees to constantly re-create their own variation of a mental image of what they think is the object of the discussion.

This can be illustrated with an example. Which of the following scenarios do you think results in a better depiction of reality?

Scenario 1 (equivalent to not using any recording device): Three people are separately shown a different boat. Then, without speaking to each other, they are asked to describe to a sketch artist what they saw.

Can you imagine how difficult it would be to get these three individuals to agree on what the boat looks like? They all think they are talking about the same boat but in fact are describing three different boats. What they don't realize is that the other two people did not see what the one person saw, and this results in many discussions.

Scenario 2 (equivalent to having a flip chart or whiteboard and a visible picture of a boat): Three individuals are asked to look at the same boat and then describe to a sketch artist what they saw.

In this case, they are more likely to become complementary and build on each other's ideas since they are all looking at the same boat.

Now that the team is set up for success, let's discuss group dynamics and how it can affect the outcomes of meetings.

25

Group Creativity Techniques

Most meeting participants split the majority of their time among discussing issues, coming up with new ideas, and choosing which alternatives to pursue.

Although the last two activities are complementary, they are typically done sequentially as the first one focuses on the generation of new ideas without taking into account how realistic and feasible they might be, and in the second one the team uses a series of tools to analyze the validity of each of those ideas, their feasibility, and their impact.

In this chapter we will discuss techniques used to generate ideas.

BRAINSTORMING

Brainstorming is the building block of a large number of group creativity techniques. This commonly used idea-generation technique gets all participants involved by establishing a judgment-free environment where participants focus on generating a large volume of ideas.

The rules of brainstorming should be reviewed at the beginning of every session, even with very experienced teams, and the facilitator should stress that the goal is to generate as many ideas as possible in a short period of time, regardless of feasibility, cost, or duration, as these will be evaluated later.

The ideal number of participants is between five and eight. Start brainstorming sessions by making sure that all members of the team are aware of the topic being discussed. Provide the team a few minutes to review the question and to ask for clarification before starting the actual group activity.

Reinforce the concept that there are no bad ideas and encourage participants to build on each other's ideas. At this stage, the team should not be stopping to discuss or evaluate ideas.

Brainstorming 101

- Agree on time allocated for brainstorming

- Post what is being brainstormed in a location visible to all participants

- Review the rules of brainstorming

- Allow participants time for individual thoughts

- Collect and post ideas as they are generated by participants

- Once the ideas stop, review the list and delete any duplicates

The facilitator is free to determine whether adding more structure—such as using a round-robin approach, where participants take turns sharing their ideas—would result in more ideas.

As the participants run out of ideas, the facilitator may choose to use the traditional categories (6Ms) of a fishbone diagram—machine, methods, material, measurement, manpower (people), and Mother Nature (environment)—to probe participants and push them to consider aspects they may have overlooked.

NOMINAL GROUP TECHNIQUE

Not everybody on a team will be equally vocal. Some brainstorming sessions have people and the managers to whom they report in the same session, and that may lead to less participation from both the junior team member in deference to the boss and the senior member in the interest of not intimidating others who report to him or her. In other cases, some participants are very shy, embarrassed, or scared to voice their opinion on some controversial topics. Your role as a facilitator is to pull out the information from them and make it available to the team.

Using the nominal group technique, the group meets formally, yet encourages independent thinking (Figure 25.1).

The nominal group technique is a brainstorming approach that encourages contributions from everybody by having each participant write down ideas and place them on the flip chart in silence. No discussions are allowed at this point of the process.

This is repeated until all ideas are exhausted and recorded on the flip chart. This is followed by discussion where ideas are clarified and reorganized into an affinity diagram.

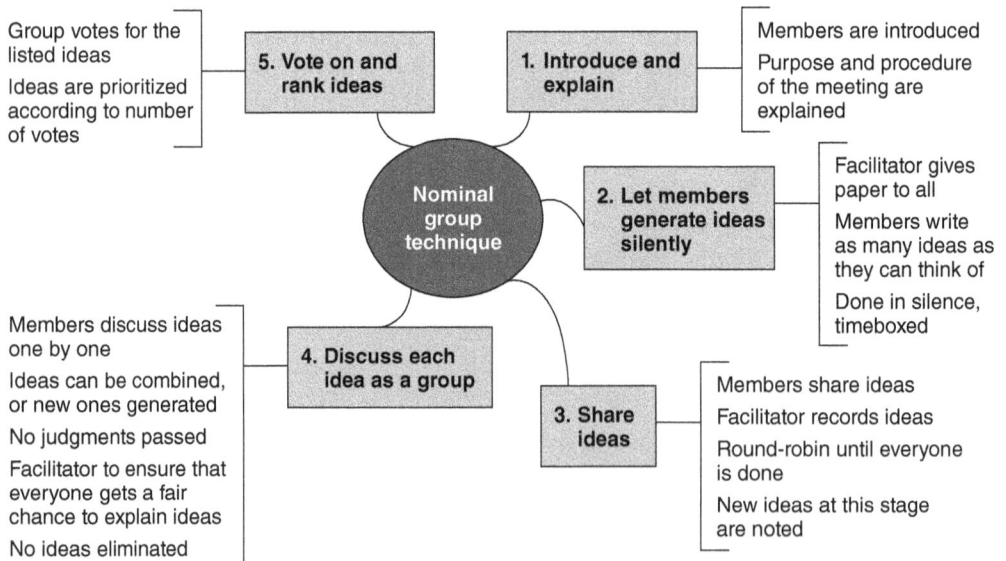

Figure 25.1   Mind map of nominal group technique.

AFFINITY DIAGRAM

An affinity diagram is a tool that leverages creativity and intuition to organize a large number of ideas into categories based on natural relationships, making them easier to handle. It is so commonly used in conjunction with brainstorming and the nominal group technique that some people think it's a step in these approaches.

The word "affinity" refers to a similarity of characteristics and implies a natural relationship. Affinity diagrams are most effective in the following situations:

- When a large number of ideas must be condensed into a more manageable number

- When the facts or thoughts are in chaos

- When the information is subjective

- When a logical organization of ideas is not as well suited as a creative organization

To make the process efficient, all ideas should be written on sticky notes so that they can easily be moved around. This means that ideas generated using other approaches, such as brainstorming or the nominal group technique, may have to be rewritten on a sticky note.

The sticky notes should be legible and written in large letters visible to all participants. The wording should be clear, concise, and to the point.

Once all the notes are gathered and displayed for all participants to see, proceed to reorganize them in silence. Place those that appear to be related next to each other, creating clusters. Notes that fit into two clusters should be duplicated and placed in both. Notes that do not fit in any of the clusters should be left as independents for now (Figure 25.2).

The team proceeds to review and discuss each cluster to determine the common theme and select a heading that represents these ideas. Some sticky notes may have to be moved to other clusters. This helps narrow a large list of possibilities into a smaller list of the top priorities or to a final selection.

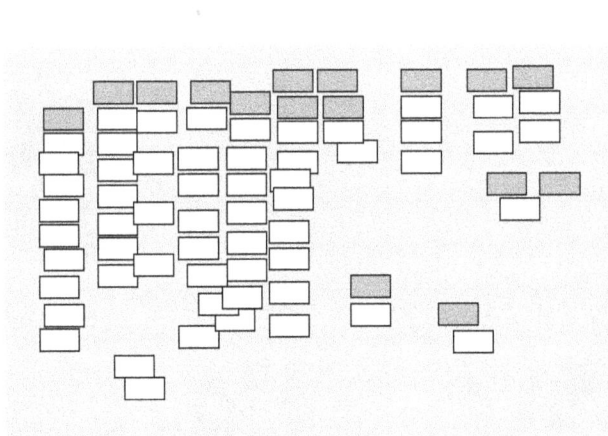

Figure 25.2 Mockup of an affinity diagram ready for review with team members.

26
Handling Difficult Group Situations

The group life cycle explains the five stages every team goes through. While experiencing challenges is normal and should be expected at any stage, working groups experience it more in the forming, storming, and norming stages.

In this chapter, we will discuss some of the challenges commonly faced by facilitators and present some techniques to help minimize the disruption and its impact on the productivity of the group.

It should be noted that some of the situations may affect only one individual or a small subgroup. Setting up the team in such a way as to keep these situations from happening in the first place is the best-case scenario.

The facilitator can always choose to either ignore them or address them when they happen.

LACK OF INTEREST, DISENGAGED TEAM, LOW PARTICIPATION

The team members seem uninterested in the topic being discussed. Their body language shows it in a variety of ways, such as looking sleepy, texting on their phones, e-mailing on their computers, daydreaming, or dozing off.

Suggested resolution approaches include the following:

- *Start with yourself.* Is your body language, voice, or attitude projecting a sense of urgency, commitment, or interest? Are you tired or feeling overwhelmed? In those situations, I have found that having someone else take over the facilitation for a short period of time gives me the opportunity to reassess the situation, catch my breath, and reconnect with my team.

- *Observe the team's body language.* Do they look bored, tired, or hungry? Do they look confused, lost, or angry?

- *Take a short break.* Prolonged and uninterrupted work sessions could lead to teammates focusing on finding ways to leave the meeting rather than focusing on the meeting itself. Taking a short break allows your participants to tend to other business or a pending personal matter or grab something to eat. Be aware that setting a meeting earlier than the beginning of the workday, too close to the lunch hour, or at the end of the workday may disrupt other aspects of the participant's life and lead to a lack of interest. Take this into consideration and schedule accordingly.

- *Ask the group what is going on and be ready to address the elephant in the room.* A special event or announcement happening inside or outside the company that you may not be aware of could be preoccupying or distracting the team. Stop the task you're working on and ask the group what is going on. It is important for you to acknowledge what is going on with the group and react accordingly, even if it means reminding the group that the event is likely out of their control and that there is a task at hand that needs to be completed. The most important thing is just to acknowledge the elephant in the room.

- *Provide descriptive feedback on what you see.* The team may have gotten complacent, not realizing that you're paying attention. By providing constructive feedback, you will give the team the opportunity to rise to the occasion.

- *Use a more active approach in starting an activity or a discussion.*

- *Ask the group if what is being discussed is working, helpful, and on track.*

- *Increase your own energy, animation, and pace.*

SIDE CONVERSATIONS

A side conversation occurs when a subset of people in a group turn to one another and speak among themselves rather than as part of the team (Figure 26.1).

Side conversations can happen both because of the lack of interest and because of too much passion. It is important in those situations not to embarrass or alienate the participants involved in the side conversation. There is a chance they are discussing a point relevant to the discussion, asking for clarification, or forming an opinion.

Figure 26.1 Side conversations.

In most cases, the situation can be addressed in a subtle way that the rest of the group won't notice. I recommend trying the following steps in order and allowing enough time in between each step for it to take effect:

- Move closer to the individuals having the side conversation
- Establish direct eye contact with the individuals
- Pause/stop speaking
- Point out that their input is important to the group and ask them if they're willing to participate
- Take a break and speak to the individuals directly in private (in extreme cases)

Establishing and maintaining a rhythm is an important aspect of facilitation: too slow and your team may lose interest, too fast and you may overwhelm some team members. In both cases, developing an agenda with predetermined time allocations, communicating it to the team, and referencing it as progress is made should allow you to gauge the rate of completion per plan. Assigning a timekeeper and establishing meeting roles, ground rules, and a decision-making mechanism should also contribute to staying on track.

GROUP GOING TOO SLOW

If the group is going too slow, you, as the facilitator, should:

- Share your observation with the group; ask if anybody else has noticed that the group is moving slowly and feels the group should move along faster
- Refer to the agenda and compare the current progress with the expected level of completion
- Try to close the agenda item being discussed
- Urge the group to move on
- Take a short break if necessary

GROUP GOING TOO FAST

If the group is going too fast, you, as the facilitator, should:

- Share your observation with the group; ask if anybody else has noticed that the group is moving quickly and feels the group should slow down
- Refer to the agenda and compare the current progress with the expected level of completion
- Review whether deliverables were completed satisfactorily
- Use the round-robin approach so that each person may voice an opinion on the topic being discussed
- Break up the team into smaller work groups and require a formalized progress report each time the team is reassembled

Figure 26.2 Disagreement.

DISAGREEMENTS

As your team progresses from the forming to the storming stage and tries to establish an operating rhythm, ideas may clash and tempers may flare (Figure 26.2).

It is important for the facilitator to be observant enough to determine when to interfere and when to let the situation run its course. It is important for the facilitator to:

- Recognize that disagreements can lead to a positive outcome

- Remind the participants of the purpose of the meeting

- Express confidence that agreement can be reached

- Encourage collaboration to find a solution

It is not uncommon for two people to say the same thing using different words, not realize it, and then fiercely defend a point of view that they both share.

Take a few minutes and have the team members determine whether they are in *violent agreement* by asking them to reword their position and to list:

- Items they agree on

- Items they disagree on

GROUPTHINK

On the other end of the spectrum from disagreements is groupthink, where making decisions as a group is favored, to the detriment of creativity and individual responsibility. Groupthink occurs when the desire to conform and reach harmony supersedes reasons and results in irrational decisions.

As a facilitator, you can address groupthink by:

- Challenging the team to review its charter and deliverables and critically evaluate the current proposals and solutions

- Having participants move and change who they work with

- Playing the role of devil's advocate

- Dividing the team into subgroups and requesting that each group come up with the most outrageous alternative solution it can think of and then present it to the rest of the team

- Asking the team to brainstorm all the reasons an idea or proposal will not work

- Bringing in a subject matter expert and/or somebody completely unfamiliar with the topic being discussed and have them challenge the team (in extreme cases)

27

Handling Difficult Individual Situations

Handling difficult situations involving one individual presents its own set of challenges and risks. Everybody has feelings, and while you may feel that tough love is the best way to go, remember that you are dealing with adults who may not respond positively to that approach.

In some cases, people behave a certain way because of a character trait; in other cases, they may have another source of motivation. In this chapter we will review certain behaviors and discuss what might motivate people to display them.

While some of these behaviors may deprive the team of the benefits stemming from the insight of one person, other behaviors describe the group and in some cases can completely derail the team and affect the contribution levels of other team members.

NONPARTICIPATION

Understanding why participants choose to stay quiet is key in this situation.

It is often difficult to distinguish between a shy individual and an individual who is scared to voice their opinion because it may be different from the apparent group consensus.

A lack of participation may also be due to the mix of participants; for example, lower-ranked employees may feel it's inappropriate to voice their opinion if their boss is present. Here are a few suggestions that may help with nonparticipation:

- Create a safe environment and make sure every participant knows their ideas will not be used to retaliate against them. This may require you to shuffle the team of participants or remove authority figures that some may fear.

- Avoid singling out one person; instead, make a general statement expressing how important it is to hear from everybody. Remind them that they each have a unique perspective and something to contribute.

- Break out into groups of two or three and have them report back.

- Talk to the person privately during a break and find out why they are not participating. Be sure to remind them that their ideas are important.

- Know that in some cases it's OK to call out participants who have contributed a lot in the past but now look distracted.

- Thank each person when they contribute. In all cases, positive reinforcement is key.

TALKING OFF TOPIC

The facilitator should keep in mind that the person talking off topic may not realize that they are doing so. It is important to be tactful as you steer them back to the current discussion. This is such a common occurrence that every facilitator should anticipate it and try to prevent it by setting up a parking lot and a rabbit hole interruption system at the beginning of the meeting.

Asking the person speaking off topic to explain the linkage between what they are saying and the current agenda is a subtle yet effective way to either make them realize they are off topic or help the other participants gain a better understanding of this person's point of view.

REPEATING THE SAME POINT

If a participant persists in repeating the same point, the facilitator should try to understand why before choosing a course of action.

More often than not, the repetition is due to a perception that their point did not come across as intended or did not receive the attention it deserves. In both cases the individual may be appeased and the meeting may move forward if the facilitator:

- Demonstrates that the point has been heard, either by repeating it or by capturing it in the parking lot

- Acknowledges the importance of the point and the person's determination

- Explains how and when the point will be dealt with

In extreme cases, the facilitator may have to propose a short break and speak to the person in private, requesting that they let it go for now.

NEGATIVE AND ANTAGONISTIC

Negative and antagonistic participants can quickly alienate other participants, create a sense of despair and/or pointlessness, and cause overall disinterest. Their behavior may be due to their ego, frustration, confusion, insecurities, or past failures with similar meetings. Tactful, firm, and immediate containment action will minimize the impact of such a behavior on the group's overall attitude toward the task at hand. When dealing with these situations, the facilitator should assess the person's willingness to collaborate by:

- Acknowledging their point of view

- Asking if there is any part of the work they feel good about

- Asking for their opinion about what is needed, recording it, and involving the other group members by asking them to respond

PERSONAL AGENDAS

Personal agendas are particularly troublesome because they imply a conscious decision to place one's selfish interest before the team's interest. Unfortunately, they are not always obvious. They are often skillfully hidden behind seemingly good intentions and must be dealt with swiftly.

Having a detailed agenda complemented by a well-written project charter is key to neutralizing personal agendas. The facilitator can:

- Ask the person how what they are saying relates to the current agenda item

- Record the point, thank the person, and move on

- Use a parking lot for off-topic issues

- Ask the person what the group should do with the input

DOMINATING

Certain individuals tend to dominate conversations that they are part of. To prevent domination, you, as the facilitator, can:

- Stop the person, thank them, and ask if anyone else has a thought on the subject

- Use inclusion activities such as nominal group technique

- Give the person a time limit or set a time limit for any one person to give feedback

- Break eye contact, move away from the person, and stop giving them focused attention

- Speak with the individual before the meeting and discuss ways to maximize group participation

Glossary

affinity diagram—A management and planning tool used to arrange large amounts of ideas and issues from multiple sources into logical categories based on the natural relationship of the items.

agenda—A list of topics or points to discuss during a meeting or workshop.

audit—A planned, independent and documented assessment to determine whether agreed-upon requirements are met.

benchmarking—To compare products, services, and processes to similar products, services, and processes of the best competitors or recognized leaders in the industry.

best practice—An established way of doing something that is accepted or prescribed as being correct or most effective.

Black Belt—A professional who is well versed in Lean Six Sigma methodology and has a thorough understanding of all aspects within the phases of DMAIC. These individuals lead improvement projects, typically in a full-time role.

bottleneck—A resource whose capacity is less than or equal to the demand placed on it and therefore limits the capacity of the overall process.

brainstorming—A formal technique used to encourage creative thinking, to generate ideas in a short period of time, and to achieve group participation in problem identification and/or solutions.

catalyst—A person or thing that causes an event or situation to happen.

champion—An executive-level manager who is responsible for managing and guiding the team for a particular initiative.

change management—A structured approach to ensure that changes are implemented thoroughly and smoothly, and that benefits are sustained.

charter—A document written to describe the work of the team and to specify objectives.

continuous improvement—A standard process used to plan, control, and improve work processes and output.

cross-functional team—A group of people with different functional expertise working toward a common goal. Typically, the team includes employees from all levels of an organization.

customer—Entity that receives a value-added product or service.

defect—(1) A quality characteristic departing from its intended level or state and sufficiently severe to cause an associated product or service to not satisfy intended normal, or reasonably foreseeable usage requirements. (2) The waste associated with producing anything that does not meet customer requirements in terms of cost, timing, or any quality specifications.

effect—Result of a cause.

essential non-value creating activities—These activities are a subset of non-value creating activities the customer is not willing to pay for, but required to meet third party expectations or requirements.

excess inventory—The waste created by any inventory not required for current customer demand.

facilitate—Comes from the Latin "facilis," for easy. It means to make something easier or more likely to happen.

facilitator—The person charged with performing all the tasks necessary to make the group's work more effective by structuring and staging meetings for success by eliminating all sources of hassle, delays and inconvenience.

failure modes and effects analysis (FMEA)—Used to assess, manage, and reduce risk associated with the failure or potential failure of products, processes, services, and other systems before they occur.

firefighting—Reactive mode.

fit—Range of mechanical interference tightness or clearance looseness of contact of mating parts when they are assembled.

flip chart—A set of sheets, hinged at the top so that they can be flipped over to easily capture ideas from a group discussion and show information or illustrations in sequence.

flow kaizen—Analysis method that considers the entire value stream to set up an action plan to create a future state.

form, fit, and function—Three attributes of an item on an ascending scale of criticality, from "form" least critical to "function" most critical which can be used to classify its defects.

form tolerance—Feature variation permitted from the perfect form specified on a drawing.

function—A task, action or activity that must be performed to achieve a desired outcome.

gemba—The place where value is created, where the work or interaction takes place.

Go Look See—The act of actually visiting the work area to gain a better understanding of what actually happens.

ground rules—Rules that are generic in nature and promote the creation of a serene and peaceful atmosphere, which encourages creativity. They take into account that people have responsibilities and needs beyond the scope of their work.

groupthink—Groupthink occurs when a homogenous, highly cohesive group is so concerned with maintaining unanimity that it fails to evaluate all its alternatives.

guideline—Documented instructions that are considered good practice but are not mandatory.

handoff map—Tool used to understand information flow and eliminate unneeded handoffs in order to minimize the number of non-value-adding steps information must flow through.

icebreaker—A mechanism by which the facilitator can help a team participant feel comfortable enough to introduce themselves and share a fact that helps the team get to know them better.

individual (point kaizen)—At the individual workstation level, there are always opportunities to reduce waste—workplace organization, inventory and tool location, work sequence, ergonomics, poka-yoke, and on and on.

inputs—Usually written as nouns and can be physical objects or information.

inspection—Activities to determine conformity, including observing, measuring, examining, testing, or gauging one or more characteristic of a product or service.

kaikaku—Japanese word used to describe a business concept concerned with making fundamental and radical changes to a production system, unlike Kaizen, which is focused on incremental minor changes.

kaizen—A Japanese word used to describe a team approach to quickly break apart and rebuild a process to function better.

kaizen blitz—An event carried out in a local area, but it involves more time and outsiders. These events address more complex issues than what the work team can comfortably handle.

kaizen kanban—A prioritized improvement display board that tells improvement teams what pre-approved projects are most relevant to current business needs and are next in line for implementation.

kanban—Japanese word for signboard or billboard implemented to signal to workers what to build next or what parts to retrieve.

key business objectives—A specific result that a person or system aims to achieve within a time frame and with available resources.

kickoff meeting—The first meeting with the project team and the client of the project.

lean—The core idea is to maximize customer value while minimizing waste. Simply, lean means creating more value for customers with fewer resources.

manufacturer—The entity which processes raw material for, or purchases hardware to be used as, components, and/or who partially or totally assembles the components into an item.

manufacturing—The organization responsible for physically producing an item.

material—A generic term for item types, including equipment, but primarily raw material or chemical materials in compounds or mixtures of any physical phase states.

measurement—The result of measuring.

methodology—A set or system of methods, principles, and rules for regulating a given discipline, as in the arts or sciences.

metric—Quantified figure of merit for useful accuracy and reliability.

nominal group technique—A brainstorming approach that encourages contribution from everybody by requiring that each participant write down ideas and place them on the board in silence.

non-value-adding activity—Any activity that consumes time or resources that your process does not require, that does not satisfy customer demands nor requirements or that customers are not willing to pay for because it does not add value to them.

observation—An instance of measuring or classifying some item being examined.

OFI—Opportunity for improvement.

outputs—Should be tangible and are usually written as nouns. They can be products, services, information, decisions, or documents.

overprocessing—The waste created by unnecessary operations and the duplication of some tasks, such as inspection.

overproduction—The waste created by using resources to assemble unneeded parts ahead of schedule while production is waiting for other needed parts—in other words, producing more than what the customer wants.

paradigm shift—An event causing everybody to go to zero, a new starting point.

parking lot—A large sheet of paper on the wall where off-topic issues are stored for future reference.

PICK chart—A Lean-Six Sigma tool used to categorize process improvement ideas during a Lean-Six Sigma event.

problem statement—A brief piece of writing used to inform the reader about the problem or issue, why it matters and why a solution should be identified as quickly and directly as possible.

process—A collection of interrelated actions, activities, steps or tasks executed to achieve a specific result per customer requirement.

process map—A visual representation of the steps required to complete a task. It displays these events sequentially using an agreed upon set of symbols and, therefore, can be used as an effective analysis and communication tool.

process operator—Person who performs the actual work necessary to achieve the objectives of that process step.

process owner—Person who has the authority to make changes and who is responsible for the performance of a process step as measured by key business indicators.

project—Planned set of interrelated tasks to be executed over a fixed period and within certain cost and other limitations.

project charter—A formal document that sets objectives, imposes constraints, and gives the authority to obtain resources and assemble the team. A project charter is a great tool used to concisely describe the work of the team and communicate the purpose of the project to all attendees what the team is trying to accomplish.

project management—A project is a temporary endeavor designed to produce a unique product, service or result with a defined beginning and end undertaken to meet unique goals and objectives.

quality—Conformance to requirements and fitness for use.

rework—Correcting of defective, failed, or non-conforming item, during or after inspection.

risk—(1) The statistical probability of attaining an unfortunate false conclusion when sampling. A good lot may be rejected or a bad lot may be accepted as a result of an unrepresentative sample. Related concepts are producer risk, buyer risk, alpha risk, and beta error. (2) The uncertainty of attaining a goal, objective, or requirement pertaining to technical, performance, cost, and schedule.

rolling action item list (RAIL)—A tool used to keep track of the actions the team is set to accomplish and their status.

scope—The extent of the area or subject matter that something deals with or to which it is relevant.

scrap—A material review disposition that releases the discrepant material for destruction. The discrepancy is so drastic that the item cannot be used or feasibly and economically corrected for future use.

seven ways—A group participation technique where participants re-create a process map seven times using sticky notes.

side conversation—Conversation that occurs when a subset of people in a group setting speak among themselves rather than as part of the team.

sponsor—A manager that supports and reviews the improvement initiative work of the team.

stakeholders—People that are affected by the project outcome or that may influence results but are not necessarily involved with project work.

standard operating procedure (SOP)—A policy and procedure document that describes the regular recurring activities appropriate to quality operations.

starburst—A symbol in the shape of an explosion that visually indicates an opportunity for improvement at a step on a process map or a value stream map.

subject matter expert—Person who knows exactly what it takes to do a particular job and is considered to be an authority on a specific subject or area.

supplier—An entity which provides a product or service, usually of value added raw material to be used as part of production, to a user or customer.

supply chain kaizen—Supply chain project teams comprise part-time representatives from participating companies within the value stream. They are focused on optimizing the entire value stream.

team—A group of people who come together to achieve a common goal.

team roster—A list of the people or things that belong to a particular group or team.

transportation waste—The waste created when transporting parts or documents from one operation to the next or to and from storage locations.

tribal knowledge—Undocumented information that is assumed to be factual, but without proof, and is handed down from one generation to the next within a group.

trigger—Causes something to happen or lets the operator know the conditions are right to proceed to the next step.

unnecessary motion—The waste created by making extra movements; it applies to the human element, not the machine element.

unused employee talent (often referred to as the eighth waste)—The waste associated with the failure to effectively engage people in the process, resulting in the under-utilization of their talents, skills, and knowledge. Any time the team fails to make the most of an employee's potential capability is a lost opportunity.

value-adding activity—Activities that change size, shape, form, fit or function of material or information to meet customer demands and requirements. Any service the customer is willing to pay for, including inspection or storage.

value stream—Sequence of activities required to design, produce, and provide a specific good or service.

variation—A change or difference in condition, amount, or level, typically with certain limits.

voice of the customer—A term that describes your customer's feedback about their experiences with and expectations for your products or services.

waiting—The waste created by keeping a resource inactive because something required to perform the task at hand is missing.

waste or muda—Any activity that consumes resources and creates no value for the customer. Any material having no measurable recovery value and considered a loss.

work instructions—A tool provided to help someone do a job correctly.

work team (mini kaizen)—A group that works in a cell or on a line segment that undertakes improvement projects affecting their collective work area.

world-class—As good as, or better than, the best competitors in the class of product competitors, as expressed in the most relevant measures and figures of merit of quality and value that can be applied to a market segment product.

Bibliography

Adair, Charlene. 2006. *Rath & Strong's Lean Pocket Guide*. Lexington, MA: Rath & Strong.

Barker, Joel. 1993. *Paradigms: The Business of Discovering the Future*. New York: Harper Collins Publishers.

Brassard, Michael, and Diane Ritter. 2010. *The Memory Jogger II, A Pocket Guide of Tools for Continuous Improvement and Effective Planning*. Salem, NH: Goal/QPC.

Dennis, Pascal. 2002. *Lean Production Simplified*. New York: Productivity Press.

———. 2006. *Getting the Right Thing Done*. Cambridge, MA: Lean Enterprise Institute.

Hutton, David W. 1994. *The Change Agents' Handbook, A Survival Guide for Quality Improvement Champions*. Milwaukee, WI: ASQC Quality Press.

Juran, J. M. 1989. *Juran on Leadership for Quality*. New York: Free Press.

Laing, Ronald David. 1960. *The Divided Self: An Existential Study in Sanity and Madness*. London: Penguin Group.

Lean Enterprise Institute. 2003. *Lean Lexicon*. Brookline, MA: Lean Enterprise Institute.

Lefever, Lee. 2013. *The Art of Explanation*. Hoboken, NJ: Wiley.

MacInnes, Richard L. 2002. *The Lean Enterprise Memory Jogger, Create Value and Eliminate Waste throughout Your Company*. Salem, NH: Goal/QPC.

Mann, David. 2010. *Creating a Lean Culture*. Boca Raton, FL: CRC Press.

Manos, Anthony, and Chad Vincent. 2012. *The Lean Handbook*. Milwaukee, WI: ASQ Quality Press.

Omdahl, Tracy. 1997. *Quality Dictionary*. W. Terre Haute, IN: Quality Council of Indiana.

Scholtes, Peter R., Brian L. Joiner, and Barbara J. Streibel. 2010. *The Team Handbook*. 3rd ed. Edison, NJ: Oriel Stat a Matrix.

Straker, David A. 1997. *Rapid Problem Solving with Post-it Notes*. New York: Da Capo Press.

Tague, Nancy. 2005. *The Quality Toolbox*. Milwaukee, WI: ASQ Quality Press.

Westcott, Russell T. 2006. *The Certified Manager of Quality/Organizational Excellence Handbook*. 3rd ed. Milwaukee, WI: ASQ Quality Press.

Wedgwood, Ian. 2006. *Lean Six Sigma: A Practitioner's Guide*. Crawfordsville, IN: Prentice Hall.

Suggested Reading

Cowley, Michael, and Ellen Domb. 1997. *Beyond Strategic Vision*. Burlington, MA: Elsevier.

Crosby, Philip B. 1995. *Quality without Tears*. New York: McGraw-Hill.

Highsmith, Jim. 2014. *Agile Project Management*. Crawfordsville, IN: Pearson Education.

Nigro, Nicholas. 2008. *The Everything Coaching and Mentoring Book*. 2nd ed. Avon, MA: Adam Media and F+W Publications.

Westcott, Russell T. 2014. *The Certified Manager of Quality/Organizational Excellence Handbook*. 4th ed. Milwaukee, WI: ASQ Quality Press.

Useful Websites

American Society for Quality, http://www.asq.org

BusinessDictionary.com, http://www.businessdictionary.com

Dictionary.com, http://www.dictionary.com

Lean Enterprise Institute, http://www.lean.org

Merriam-Webster, http://www.merriam-webster.com

Vertex42, http://www.vertex42.com

Vocabulary.com, http://www.Vocabulary.com

Index

Note: Page numbers followed by *f* or *t* refer to figures or tables, respectively.

www.ingramcontent.com/pod-product-compliance
Lightning Source LLC
Chambersburg PA
CBHW081109220326
41598CB00038B/7281